Contents

Part A The Chemistry of Energy

1 Fossil Fuels   *2*

2 Nuclear Energy   *30*

3 Alternative Energy Supplies   *47*

Part B The Chemistry of Materials

4 Chemicals from Oil   *56*

5 Mankind and Metals   *77*

6 Iron and Steel   *84*

7 Aluminium and Other Metals   *99*

8 Chemicals from Salt   *127*

9 Explaining the Properties of Chemicals   *141*

Part C The Chemistry of Food Production

10 Food Production   *150*

Part D Chemistry at Home

11 Water Supplies   *178*

12 Household Chemistry   *188*

Projects

1 *Lead in the Environment*   *208*
2 *The Oak Ridge questionnaire*   *209*
3 *Anaesthetics*   *210*
4 *Chemistry in space*   *212*
5 *Choosing aluminium alloys*   *214*
6 *Chemical warfare*   *214*
7 *Spare part surgery*   *216*
8 *The Seveso disaster*   *217*
9 *Fluoridation of water supplies*   *218*
10 *Common chemicals*   *221*

Appendix 1   Chemical Formulas   *223*
Appendix 2   Chemical Equations and Calculations   *226*
Dictionary of 100 Chemical Terms   *231*

Index   *235*

Preface

I am indebted to a great many people after writing a book of this nature. Whilst I hope that the overall approach is original, I have inevitably been influenced by other writers and by many colleagues and pupils.

There are several individuals whom I would like particularly to acknowledge:

Terry Allsop, for his constructive help and encouragement throughout this project.

John Collyer, my former boss at Newmarket Upper School, for frequently bringing me down to earth.

Kath Dodd, for translating my writing into a manuscript.

Nicky, my wife, for continuous support and valuable suggestions at all stages in the production of this book.

I would also like to thank John Cushion of Pitman Publishing Ltd both for providing the initial impetus and for his unwavering enthusiasm and care.

A major part of the research for this book involved writing to numerous chemical companies and to other private and public organizations for specific items of information. I fear that there are too many to mention individually, but I would like to thank them all for their unfailing help and courtesy. I hope that I have managed to convey some of the interest and enthusiasm which they communicated to me when describing their activities.

The following people or organizations kindly supplied illustrations:

Anglian Water Authority   *Figure 3.3*
Austin Rover Group   *1.15*
BIP Chemicals Ltd   *4.13*
Britain/Israel Public Affairs Committee   *10.2*
British Alcan Aluminium   *5.7, 7.1, 7.3, 7.14*
British Gas   *1.5a, 1.5b, 1.6, 1.10*
British Leyland   *6.10*
British Museum   *5.2*
British Oxygen Company   *8.9, cover*
British Petroleum   *1.9, 4.1a*
British Rail   *7.22*
British Steel   *6.2b, 6.5a, 6.6*
Building Research Establishment   *12.1*
Bundesbildstelle Bonn   *1.22b*
J-Christo & H Shunk   *page 55*
Chubb Fire Security Ltd   *1.26*
De Beers Industrial Division   *9.7a*
Department of Energy   *1.7, 4.15*
Dunlop Ltd   *4.16*
Earthscan   *3.2, 11.3* (Mark Edwards), *11.2* (Mohamed Amin), *3.4a* (UNEP/da Silva)
Glass Manufacturers Association   *8.13*
ICI Agricultural Division   *10.8, 10.9, 10.13, page 149*
ICI Mond Division   *8.1, 8.3*
ICI Nobel's Explosives Company Ltd   *page 1 (top)*
ICI Paints Division   *4.17*
ICI Plant Protection Division   *10.20*
Ideal Homes Magazine   *page 177*
Imperial War Museum   *2.1b*
Ind Coope Ltd   *12.3*
Magnesium Elektron Ltd   *7.30*
Materials Reclamation Weekly   *6.9*
National Aeronautics and Space Administration   *6.1, page 1 (bottom)*
National Coal Board   *1.16, 1.17, 1.18a, 1.18b*
New Scientist   *3.5* (Robert Lawrence)
North of Scotland Hydro-Electric Board   *3.6*
Permutit-Boby Ltd   *11.6*
Pete Addis   *1.22a*
Pilkingtons Ltd   *8.12*
Rexel Ltd   *9.7b*
Rio Tinto Zinc Ltd   *7.24, 7.25, 7.26, 8.2*
Shell   *1.12a, 4.1b, 4.7a, 4.10, 12.9*
Sinclair Research Ltd   *12.16*
South African Embassy   *6.8*
Steve Venables   *4.14*
Terry Jennings   *12.13*
Titanium Metal and Alloys Ltd   *6.7*
United Kingdom Atomic Energy Authority   *2.1a, 2.10a, 2.10b, 2.11a, 2.11b, 2.13a, 2.13b*
United Kingdom International Solar Energy Society   *3.8a*
Waitrose Ltd   *12.7*
Zinc Development Association   *7.15*

Chemistry in Use

Dr Roland Jackson
Head of Chemistry
Backwell School
Avon

with advisory participation from
Terry Allsop
Lecturer in Education
University of Oxford

Pitman

PITMAN PUBLISHING LIMITED
128 Long Acre, London WC2E 9AN

Associated Companies
Pitman Publishing New Zealand Ltd, Wellington
Pitman Publishing Pty Ltd, Melbourne

© W R C Jackson & R T Allsop

First published in Great Britain 1984

All rights reserved. No part of this publication may be reproduced, stored in a retrieval system, or transmitted, in any form or by any means, electronic, mechanical, photocopying, recording and/or otherwise without the prior written permission of the publishers. This book may not be lent, resold, hired out or otherwise disposed of by way of trade in any form of binding or cover other than that in which it is published, without the prior consent of the publishers.

ISBN 0 273 01981 3

Printed in Great Britain at The Pitman Press, Bath

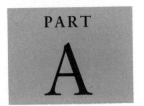

PART A

The Chemistry of Energy

We cannot live without energy. We need it for everything we do. Without supplies of energy we could not build houses or factories, keep ourselves warm, make tools or other objects, transport ourselves or our goods, grow crops, cook food, provide clean water, or treat sewage.

Many chemicals are useful stores of energy. Almost all the energy used in the world today comes from a group of chemicals called fossil fuels. This section starts with these important chemicals.

Explosives in action. Energy is given out violently by some chemical reactions.

A Saturn V space rocket carrying the Apollo 10 moon mission. The energy used to power the rocket comes from burning the fossil fuel called kerosene.

Fossil Fuels 1

1 Fossil Fuels

What are Fossil Fuels?

There are three main **fossil fuels**—natural gas, oil and coal. These chemicals are called fossil fuels because they were made millions of years ago from the remains of animals and plants.

In the U.K. about 95% of the energy which we use comes from fossil fuels (*figure 1.1*). Our way of life depends on them.

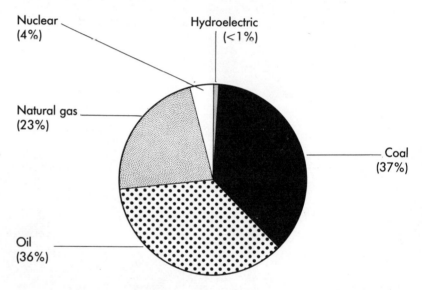

Figure 1.1 The three fossil fuels, coal, oil and natural gas, provide most of the energy used in the U.K. (1981).

The chemicals in fossil fuels

The three types of fossil fuel look completely different but they all contain the same sort of chemicals, called **hydrocarbons**. As the name suggests, hydrocarbons are chemical compounds which contain only hydrogen and carbon. When hydrocarbons burn in air, they give out energy. This energy can be used for many purposes. For example, we use it to heat our houses, run our factories, and power our transport.

Formation of natural gas and oil

These chemicals are often found together. They were probably formed from small sea creatures which sank to the sea bed when they died. The dead creatures were slowly covered by mud and sand, while their bodies rotted to form the gas and oil.

Most supplies of gas and oil are now hundreds or thousands of metres below the surface of the earth. Deep wells have to be drilled to get them out (*figure 1.2*).

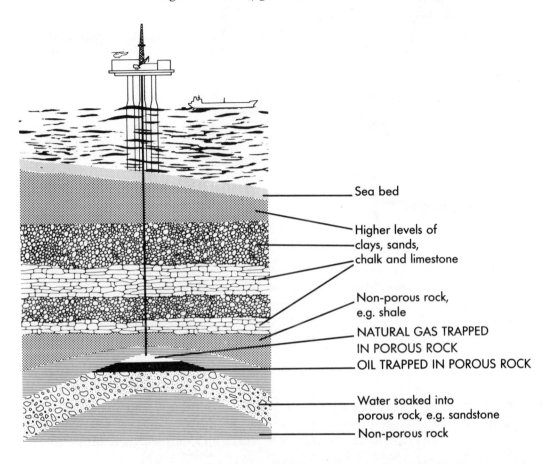

Figure 1.2 Natural gas and oil underground. The oil is soaked into a porous rock like sandstone. (A porous rock contains many holes which a liquid like oil can enter.) The oil is trapped between layers of non-porous rocks, which it cannot soak through. Gas is often found above the oil. Modern gas and oil wells can be 8 km deep.

Formation of coal

In the carboniferous period, about 300 million years ago, the land was covered with trees and other plants. When the plants died, they were washed into swampy areas. They started to decay, forming peat. The peat was then covered by layers of mud and sand. Over millions of years, the peat was compressed to form coal, while the mud and sand became shale and sandstone.

The coal is found today in layers called seams (*figure 1.3*).

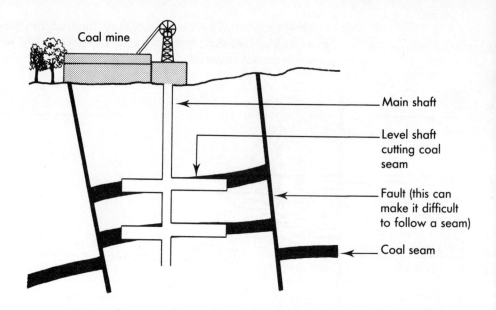

Figure 1.3 Coal underground. The coal is found in thin layers called seams, usually 1–2 m thick. If the seam is near the surface, open-cast mining can be used to get it out. If the seam is underground, a main shaft is driven down and level shafts are cut into the seams. Modern mines can be 1200 m deep. Care has to be taken to keep the mine ventilated and dry, and to prevent explosions of gas or coal dust.

The energy stored in fossil fuels

Fossil fuels are really stores of **solar energy**, built up over many years. Energy from the sun is needed for plants to grow. Without this solar energy there would be no plants and therefore no coal (*figure 1.4*).

Animals do not use solar energy directly like plants, but animals could not live unless plants were there for them to eat. In order to grow, animals make use of the solar energy stored in plants. This means that natural gas and oil, formed from dead sea animals, are also stores of solar energy. The stored solar energy is released in a useful form when the fossil fuel is burnt.

Natural Gas
Finding natural gas

It is an expensive business to find natural gas and to get it from the ground. Scientists first have to decide where any gas might be found. Their surveys may take several years. These surveys could include aerial photographs and measurement of the way in which sound waves from controlled explosions travel through different rocks below the ground. Rock samples may be taken to help the scientists to build up a picture of the rocks beneath the surface.

Finally, a decision is taken to drill a well. This can cost several million pounds and the well may not even contain any gas.

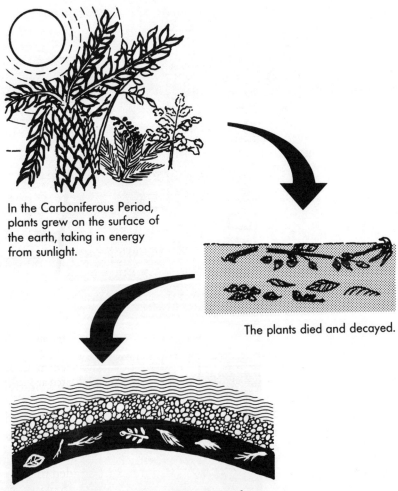

In the Carboniferous Period, plants grew on the surface of the earth, taking in energy from sunlight.

The plants died and decayed.

Coal was produced by slow compression of the plant remains. The prints of fossilized leaves and wood can be seen in lumps of coal today.

Figure 1.4 Coal and other fossil fuels are stores of solar energy.

Most of the natural gas discovered so far is in the U.S.S.R., the Middle East and the U.S.A.

The supply of natural gas in the U.K. comes from rocks under the North Sea (*figure 1.5a, b*). North Sea gas is carried ashore through pipes. It is then distributed round the country by British Gas (*figure 1.6*). At present, there are about 230 000 km of gas pipes in the U.K.

Using natural gas

Most natural gas is burnt as a fuel. Natural gas supplies about one third of all energy to British industry and is used to heat half the nation's houses (*figure 1.7*).

Figure 1.5a The Britannia rig, used by British Gas in the North Sea.

Figure 1.5b The drilling team at work on the Britannia rig.

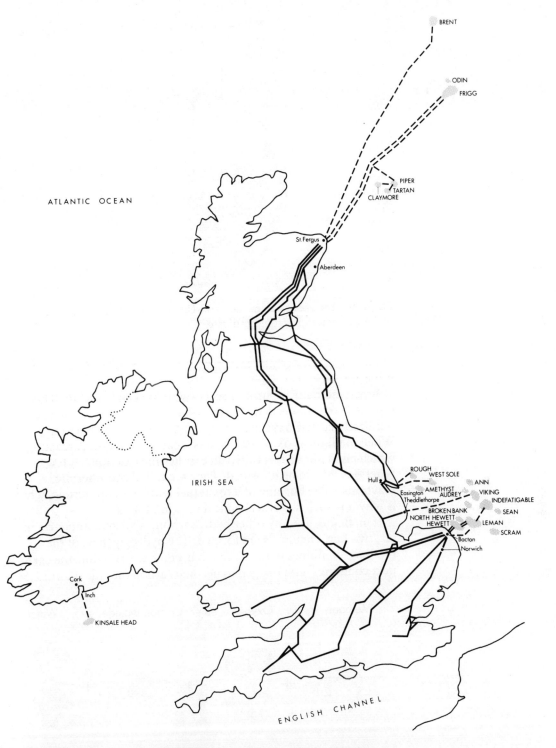

Figure 1.6 Map of U.K. gas wells and pipelines.

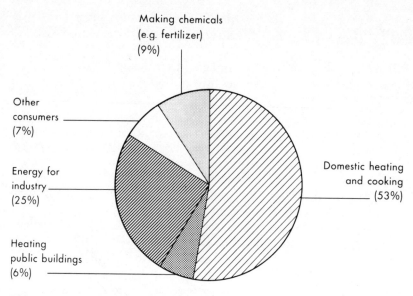

Figure 1.7 The use of natural gas in the U.K. (1981). The amount used for making chemicals is a rough estimate.

Natural gas is mostly *methane*, which is the simplest hydrocarbon. It has the formula CH_4.

When methane is burnt, it reacts with oxygen in the air. This type of reaction is called an **oxidation**. The methane is said to be oxidized because oxygen is combining with it in the reaction. When methane is oxidized like this, heat is given out. A reaction which gives out heat is called an **exothermic** reaction. (A few reactions take in heat instead. They are called **endothermic** reactions.) The burning of a fossil fuel is always exothermic, which is why it is such a useful chemical reaction for us.

If methane, like any other hydrocarbon, is burnt completely in air, two new chemicals are formed. The hydrogen part of the methane combines with oxygen to become water, while the carbon part combines with oxygen to become carbon dioxide (*figure 1.8*).

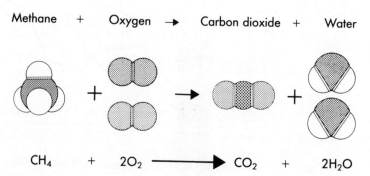

Figure 1.8 Methane burns when it reacts with oxygen in the air. Carbon dioxide and water are made if the methane burns completely.

Some natural gas is not burnt, but is used in the chemical industry. Fertilizers are made from natural gas in the U.K. (p. 155).

Problems with natural gas

Natural gas is a convenient fuel. It burns with a clean flame and causes little pollution. We use natural gas so much that we take it for granted. We should not do so. Natural gas is a fossil fuel. There is only a certain amount of it in the ground. Once it is taken out and used, it cannot be replaced. We say that the amount of natural gas is *finite*, which means that there is only a limited amount available.

Scientists calculate that the whole world's supply of natural gas may only last for another 50 years or so. In our lifetime we may have an *energy crisis*. We will probably go back to making gas from coal. There are already plans to do this, but coal itself will not last for ever. In the meantime, we should obviously try to use the natural gas as slowly as possible. Insulating houses is one way of conserving this precious gas. Not only does it save on the heating bill, but it also makes the gas last longer.

Crude Oil
Finding crude oil

Crude oil is found in the same way as natural gas, since they are often together in the same rocks. Many oil wells are on the land (*figure 1.9*), although exploration has also moved to the sea.

Figure 1.9 An oil well in Alaska.

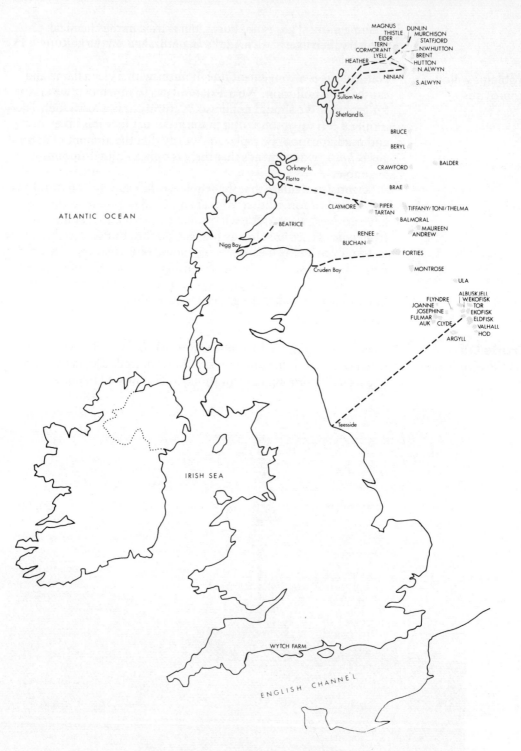

Figure 1.10 Map of U.K. oil wells and pipelines.

There are four major oil-producing areas in the world. These are the Middle East, the U.S.A., the U.S.S.R., and countries round the Caribbean (e.g. Venezuela and Mexico). Over half the world's known reserves of oil are in the Middle East.

The countries in the Middle East, especially Saudi Arabia, have become very important because they produce so much of the oil which the Western world uses. These countries, together with some in Africa and South America, belong to an organization called OPEC (Organization of Petroleum Exporting Countries). OPEC countries produce half the world's oil. You will often hear OPEC mentioned in the news, because it can raise or lower the price we pay for oil.

Crude oil has been found in the U.K. section of the North Sea. The oil is taken ashore to refineries by pipes or tankers (*figure 1.10*). The U.K. now produces as much oil as it uses.

The chemicals in crude oil

Crude oil itself is not much use, because it is a mixture of many different hydrocarbons. Most of the hydrocarbons in oil belong to a family of chemicals called the **alkanes**.

Alkanes

The simplest alkane is methane CH_4 which makes up most of natural gas. A molecule of methane contains one carbon atom joined to four hydrogen atoms. A diagram of this is called the **structural formula** of methane (*figure 1.11*).

Figure 1.11 The structural formula of methane CH_4, the simplest alkane.

The next members of the alkane family are ethane C_2H_6, propane C_3H_8, and butane C_4H_{10}. They are all burnt as fuels, just like methane. Calor gas contains mostly butane.

Most of the alkanes found in crude oil have much longer chains of carbon atoms. Octane is one example. Molecules of octane contain eight carbon atoms. Octane is one of the alkanes in petrol. The octane number of a petrol is used to show how smoothly the petrol will burn. Four star petrol has a higher octane number than two star petrol.

Information about some of the alkanes is given in *table 1.1*.

Name	Molecular formula	Boiling point (°C)	State at room temperature
Methane	CH_4	−161	Gas
Ethane	C_2H_6	−89	Gas
Propane	C_3H_8	−42	Gas
Butane	C_4H_{10}	0	Gas
Octane	C_8H_{18}	126	Liquid

Table 1.1 Alkanes

The way in which atoms like carbon and hydrogen join together to form molecules like alkanes is explained in Chapter 4 (p.59).

Refining oil

In an oil refinery, the mixture of hydrocarbons in crude oil is separated into smaller groups which are more useful. The hydrocarbons can be separated because they have different boiling points. In general, the smaller the hydrocarbon, the lower its boiling point (see *table 1.1*). This is why the smallest hydrocarbons, including methane, are gases at room temperature. These small hydrocarbons can still be found in crude oil because they dissolve in it.

The different hydrocarbons are separated by a method called **fractional distillation** or *fractionation*. The crude oil is heated to about 400°C and then pumped into a fractionating column, which may be 45 m high (*figure 1.12a, b*).

Figure 1.12a A fractionation unit at the Stanlow refinery in the U.K. The big column on the right separates the crude oil into four fractions (see *figure 1.12b*). The residue from this column is fractionated again, in the smaller column in the centre of the picture. Behind these columns is the bottom of a huge chimney, 143 m high. Waste gases are sent through it high into the atmosphere, so that they cause as little pollution as possible.

Most of the hydrocarbons are turned into gases by the heating, and they then start to rise up the column. As they rise up the column they cool down, because they are moving further from the heaters.

Near the bottom of the column it is still quite hot. This means that only hydrocarbons with high boiling points will *condense* (turn back to liquids). Hydrocarbons with lower boiling points will still be gases, so they carry on up the column. Further up the column it is cooler. Here, hydrocarbons with lower boiling points will condense. A few of the hydrocarbons do not condense at all, because it is never cold enough. These are the petroleum gases like methane and ethane which are dissolved in the crude oil. They come out of the top of the column.

In the first fractionation, the crude oil is usually separated into four different parts. Each collection of hydrocarbons is called a *fraction*, because it is a part of the original crude oil. These fractions can be distilled again to separate them further. In total, about eight fractions are usually produced. Information about them is given in *table 1.2*.

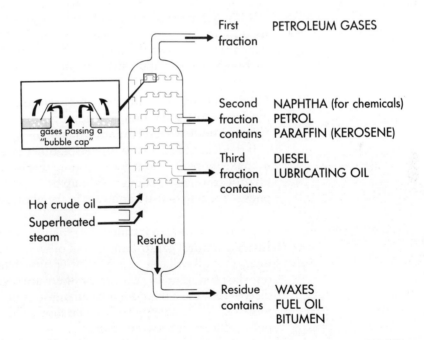

Figure 1.12b Fractionation of crude oil.
  In the first stage, the crude oil is often separated into four fractions. Hydrocarbons with very low boiling points reach the top of the column, which may be 45 m high. Hydrocarbons with medium boiling points condense in the middle of the column. Hydrocarbons with the highest boiling points form a residue at the bottom.
  The fractions which are obtained in the first stage can be fractionated again or altered further to make substances like petrol and waxes.

Fraction	Approximate size of hydrocarbons	Uses
Petroleum gas	1–4 carbon atoms	Bottled gas, making plastics
Petrol/naphtha	4–12 carbon atoms	Car fuel, making chemicals
Kerosene	9–16 carbon atoms	Jet fuel, lighting and heating houses
Diesel	15–25 carbon atoms	Diesel fuel
Lubricating oils	20–30+ carbon atoms	Lubrication
Waxes	20–30+ carbon atoms	Candles, polishes, ointments
Fuel oil	20+ carbon atoms	Fuel for ships, factories and central heating
Bitumen	largest hydrocarbons	Road surfaces, roofing

Table 1.2 Some fractions from crude oil

Using the fractions from crude oil

Most of the hydrocarbons from crude oil are burnt as fuels. For example, petrol is burnt in cars, diesel in lorries, and fuel oil in ships. Apart from powering vehicles, much of the fuel is burnt for heating or for generating electricity (*figure 1.13*).

Although most of the world's oil is simply burnt, about 10% is used for making chemicals like plastics. Our present way of life depends so much on the chemicals which are made from oil that there is a whole chapter about them (Chapter 4).

Different crude oils

Crude oil from different parts of the world contains different amounts of each fraction. The amounts of different fractions in a typical Middle East oil are shown in *figure 1.14*.

If there is a big demand for petrol, other fractions can be altered to make more of it. One way of doing this is called **cracking**. Hydrocarbon molecules are split up or "cracked" to form different ones. Cracking is also important in the plastics industry (p.64).

Problems with oil

Oil is a fossil fuel, so it cannot be replaced when it has been used. Just like natural gas, the amount of oil in the earth is finite. It is likely that oil will last an even shorter time than natural gas. It may only last for 30 or 40 years unless there are several big new discoveries. Even if it does not run out so soon, it may get more and more expensive as supplies become short. This could lead to political tension between countries.

There are also chemical problems with oil. One of these is the pollution of the sea by crude oil. So much oil is carried by sea that some pollution seems unavoidable. Accidents to oil tankers make headline news, because whole coasts can be polluted. This causes great damage to bird and fish life and sometimes to tourism. The methods of cleaning up oil spills have improved since the Torrey Canyon disaster of 1967. On that occasion, the detergent used to clean up the oil caused damage to shore life. Although accidents

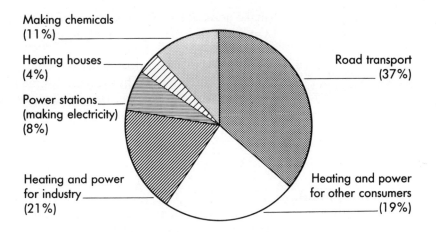

Figure 1.13 The use of crude oil fractions in the U.K. (1981).

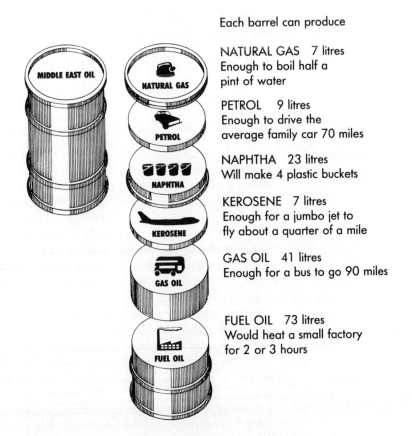

Figure 1.14 A barrel of oil contains about 160 litres. It costs around $30 (1984), but the price can go up or down very quickly.

Fossil Fuels 15

make news, they only make up a small part of oil pollution. Operations like cleaning out ships' tanks at sea cause far more pollution.

A second problem arises when oil is burnt. When any hydrocarbon is burnt, chemicals are formed which pollute the earth and the atmosphere. The chemistry of the car engine illustrates these problems.

Chemistry inside a Car Engine

The energy used to move a car comes from burning petrol. A mixture of petrol and air is taken into the cylinder of the engine. It is exploded by a spark and the explosion forces the pistons to move. The movement of the pistons then moves the wheels of the car via the crankshaft.

The amount of petrol and air entering the cylinder is controlled by the carburettor. When the accelerator is pressed, more air rushes into the carburettor. More petrol is taken in with it, so the car goes faster (*figure 1.15*).

Petrol is a hydrocarbon, so carbon dioxide and water vapour are made when it burns. These gases come out in the exhaust. Unfortunately, they are not the only chemicals produced when petrol burns.

Pollution from cars

Petrol does not burn completely in a car engine. Although most of the carbon in the fuel is turned into carbon dioxide, some of the poisonous gas carbon monoxide is made as well. Carbon monoxide combines with haemoglobin, the blood's oxygen carrier. Death can occur from lack of oxygen when carbon monoxide builds up in an enclosed space like a garage, although it would never be this serious in the open air. Particles of carbon, which we call soot, are also formed. The car manufacturers design engines to reduce these problems as much as possible, but they can never remove them completely.

There are more difficulties. The temperature inside a car engine can reach 2500°C. This is enough to make nitrogen in the air react with oxygen. Poisonous and acid gases called nitrogen oxides are made. In addition, fossil fuels always contain a small amount of sulphur. When this burns, another poisonous and acid gas, sulphur dioxide, is made.

The last big problem is man-made. Lead compounds are often added to petrol to help it to burn properly. Lead is another poison. Many countries have taken steps to remove lead from petrol. You can read more about the arguments on p.208.

Figure 1.15 Inside the bonnet of the MG Metro Turbo.

Figure 1.16 Map of the main U.K. coalfields.

Coal

Finding coal

Coal is the oldest known fossil fuel. It was certainly used in Roman times, 2000 years ago.

The largest producers of coal today are the U.S.S.R., the U.S.A., China, Poland and the U.K. It has been said that Britain is an island built on coal. There is some truth in this (*figure 1.16*).

Both types of mining, underground and open-cast, are used in the U.K. (*figure 1.17*, *figure 1.18*). The new Selby mine in Yorkshire will be the biggest deep mine in the world when it is fully developed.

Using coal

About 65% of the coal mined in the world is used to make electricity. In the U.K. the percentage is even higher (*figure 1.19*). The coal is burned in large power stations. The heat energy from this exothermic reaction is made to drive turbines, which produce the electricity.

Coal, like crude oil, is a mixture of many different substances. The substances can be separated by heating the coal without air, so that it does not burn. There are four main products or fractions (*figure 1.20*).

Figure 1.17 The surface buildings at Houghton Main Colliery in the Barnsley area. Eight miles (13 km) of tunnels connect this and two other pits with the main Grimethorpe Colliery. The colliery's buildings and pit-head baths are all heated by methane gas from the underground coal seams.

Fossil Fuels 19

Figure 1.18a Open-cast mining at Shipley in Derbyshire. 1.5 million tonnes of coal were taken from this mine.

Figure 1.18b Shipley Lake was created after the mining had finished. The National Coal Board won an award for developing this area for wildlife and recreation.

The most important of these chemicals today is coke. Coke is used in the iron and steel industry (p.86). Many chemicals, like plastics, detergents, pesticides, and fertilizers, used to be made mostly from coal. They are now made mainly from natural gas and oil. This is because gas and oil are cheaper at the moment. We may have to turn to coal again as the oil runs out.

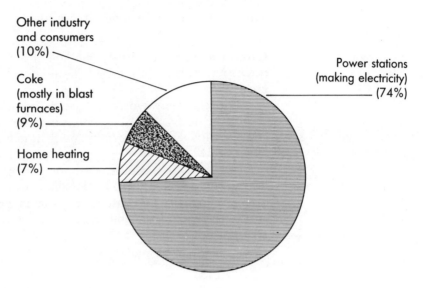

Figure 1.19 The use of coal in the U.K. (1981).

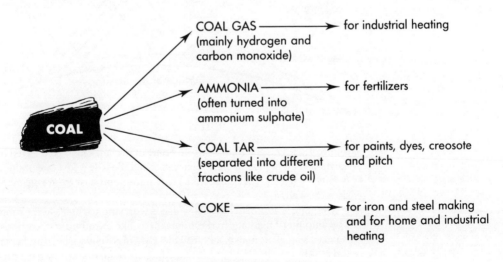

Figure 1.20 Some chemicals which are made today from coal.

Fossil Fuels 21

Problems with coal

There is much more coal in this planet than natural gas or oil. Even so, coal will eventually run out because it is a fossil fuel. At present, people think that it should last for about 250 years.

Burning coal pollutes the air in the same way that burning oil does. Governments are well aware of this. In the U.K., industries have to obey the Clean Air Act. Factories and power stations are fitted with cleaners to remove some of the pollutants. Many parts of the country are "smokeless zones", and certain types of coal may not be burnt by anyone in those zones.

The greenhouse effect

Carbon dioxide is not usually thought of as a pollutant. After all, it occurs naturally in the atmosphere and plants need it in order to grow.

Scientists are now worried, however, because the amount of carbon dioxide is increasing. This is because we are burning more and more fossil fuels, especially coal. The extra carbon dioxide may cause the earth to overheat (*figure 1.21*).

A rise of only 1°C could have serious effects. Part of the antarctic ice-cap could melt. The sea level might rise by 10 m. Would your house be flooded by this rise? If not, just imagine how many would. An even smaller rise in temperature could devastate crop-growing areas.

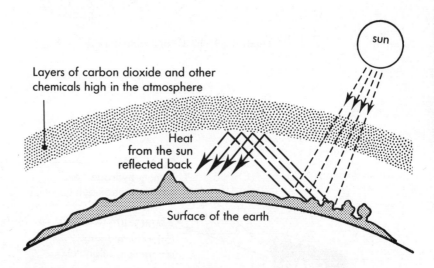

Figure 1.21 The greenhouse effect. Some of the sun's heat is reflected back to earth instead of escaping into space. Carbon dioxide is just one of the chemicals which reflects back the heat. Smoke and dust from volcanoes may do so as well.

Methane is another chemical which is worrying scientists at the moment. The amount of methane in the atmosphere, like the amount of carbon dioxide, is increasing. This may be because there are more cows, sheep and goats today. These animals all produce methane. The methane may have caused an overheating of 0.25°C, which is enough to worry about.

Fossil Fuel Pollution

The acid rain problem

Ten billion fir trees in Germany are dying.
Buildings are slowly crumbling in many parts of Europe.
Fish life is destroyed in 70% of Norway's rivers.
20 000 Swedish lakes are dead or dying.
Near Katowice, in Poland, trains can only run at 40 m.p.h. because the rails are rusted.

All this is happening just because it rains, but it is not ordinary rain. The rain which falls on most of Europe and America today is acidic. In one storm in West Virginia it was even sourer than lemon juice. The acid rain falls on the U.K. as well. One storm near Edinburgh was 500 times more acidic than normal.

The chemistry of acid rain

All rain water contains a little acid. This is because carbon dioxide from the air dissolves in the rain. An acid is made whenever the oxide of a non-metal, like carbon, is dissolved in water. The acid formed from carbon dioxide is called carbonic acid. There is only a little carbon dioxide in the air, and only a little carbonic acid is formed. The carbonic acid is useful to plants and trees because it helps them to get chemicals from the soil.

Figure 1.22*a* A stone statue outside Cologne cathedral, eaten away by acid rain.

Figure 1.22*b* The effect of acid rain on a tree in West Germany.

It is the *extra* acid which causes so much trouble today. It is made because we burn fossil fuels. All fossil fuels contain some sulphur. When the fuel is burnt, sulphur dioxide gas is formed. The sulphur dioxide dissolves in the rain and makes it more acidic, because sulphur dioxide is another non-metal oxide. It can turn rain water into dilute sulphuric acid. Other acidic gases are also made when fossil fuels are burnt. An example is nitrogen dioxide. It dissolves in rain to form dilute nitric acid. Rain water today contains a mixture of sulphuric and nitric acids.

Sulphur dioxide pollution

Sulphur dioxide is the main pollutant from burning fossil fuels. Enough sulphur dioxide passes into the air each year to give everyone in Europe about 25 kg of sulphur. More sulphur falls from the skies onto the U.K. than is used in the whole U.K. chemical industry.

Some sulphur dioxide is always found in the air. It comes from volcanoes and decaying plants and animals. Near big cities in Europe or America it is common to find ten or one hundred times the normal amount (*figure 1.23*). It can cause breathing problems to people, especially those who suffer from bronchitis.

In the U.K. the pollution comes mainly from power stations and from industries which burn coal and oil. The sulphur dioxide is released into the air through tall chimneys. It is carried away by the wind and then washed out in the rain.

Reducing the pollution

It is always expensive to stop pollution but the cost of the pollution may be even greater.

There are two main ways to reduce the pollution, apart from using less fossil fuel. One way is to remove the sulphur from the fuel and the other is to catch the sulphur dioxide before it leaves the chimney. Some sulphur is always removed from fuels like oil and gas before they are supplied to us. It is possible to remove more, but the fuels would then be more expensive. Some of the sulphur dioxide is also caught as it leaves chimneys of power stations or factories. Again, more could be caught, but it would be more expensive.

The cost of reducing pollution is not all that we need to worry about. Can we really let something like the Black Forest in Germany die, or should we take action regardless of cost?

Other Fossil Fuels
Tar sands and oil shales

About 90% of the world's reserves of fossil fuels are found in deposits of natural gas, crude oil and coal. There are two other sources of fuel: tar sands and oil shales (*figure 1.24*).

At the moment they are not worth mining, because a large amount of sand or shale has to be mined for a small amount of fuel. As natural gas and oil start to run out, these sources may become useful.

Figure 1.23 An estimate of sulphur dioxide (SO$_2$) pollution over the U.K. The lines are similar to contour lines on a map—the higher the number, the higher the pollution.

Fossil Fuels 25

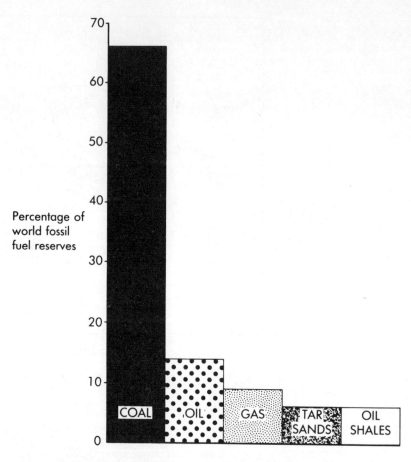

Figure 1.24 Estimated world reserves of fossil fuels, World Energy Conference, 1980.

Peat

Peat is almost a fossil fuel. It is formed from plants which have decayed for many years. Peat is quite common in wet mountain areas like the Highlands of Scotland. It can be dug out and dried to use as a fuel.

Peat can be burnt in a power station to make electricity. Making electricity in this way is as cheap as using coal or oil. There are power stations in Holland and Ireland which burn peat. Power stations burning peat can be especially useful in remote places near peat supplies.

Fires and Fire-fighting

We use fires every day, but when fires get out of control they can be dangerous and terrifying. In 1666 most of London was destroyed by fire. Imagine that happening today.

Three things are needed for a fire to burn—fuel, oxygen and heat (*figure 1.25*). If any one of these is removed, the fire goes out.

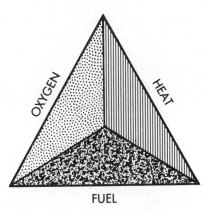

Figure 1.25 The fire triangle.
All fires need a fuel to burn. They need heat to keep the fuel alight and they need oxygen to react with the fuel. If one of these can be removed, the fire will go out. When water is used to put out a fire, which part of the triangle does it remove?

Types of fire

Fire-fighters divide fires into four types:

CLASS A: ordinary materials like wood, paper, cloth and plastics.
CLASS B: flammable liquids like cooking fat, oil and petrol.
CLASS C: flammable gases like natural gas.
CLASS D: metals.

Extinguishing fires

Class A fires (e.g. wood or paper) can be put out most easily by using water. The water cools the fire until it is no longer hot enough to burn.

Class B and Class C fires must be fought differently. Never, for example, try to put out a chip-pan fire with water. The burning fat will float on the water. Some of it will also fly into the air and may start new fires. A quick way of dealing with a chip-pan fire is to cover it with a bread-board or something similar. This will stop oxygen getting to the fire so that the fire will burn out.

A carbon dioxide extinguisher could be used instead. Carbon dioxide gas is forced out of an extinguisher under pressure. It forms a blanket over the fire which stops oxygen getting in.

Fires of oil and gas are especially dangerous in the oil industry (*figure 1.26*). Foam extinguishers or dry powder extinguishers can be used to fight these types of fire. Dry powder extinguishers are particularly good. They do not just form a blanket over the fire, like foam extinguishers, but they work right through the whole fire to stop the burning reaction. Different chemicals can be used in these extinguishers. A common example is sodium hydrogencarbonate. Just 20 kg of the best dry powder can snuff out the flaring oil or gas on a drilling rig.

Class D fires, involving metals, can be very difficult to fight. This is another problem on oil rigs. Oil fires burn at 1200°C, but steel collapses at 540°C. Special fire-resistant coatings are made to cover the steel. Hot metals can react with steam or carbon dioxide, but some dry powder extinguishers can be used. If the fire involves electrical equipment, it is essential not to use water or foam because of the risk of shocks. Carbon dioxide or a suitable liquid extinguisher could be used instead (*Table 1.3*).

Fire extinguisher	Suitable fire	How it works
Water	Paper, wood, fabric	Cools it down so the fire no longer burns
Carbon dioxide	Petrol, electrical equipment	Stops oxygen getting to the fire
Dry powder	Oil, gas	Stops the burning reaction
Liquid (Halon)	Electrical equipment	Stops the burning reaction
Foam	Oil	Forms a layer to stop oxygen getting in and provides some cooling

Table 1.3 Using fire extinguishers

Figure 1.26 Fighting a small oil spill fire at a refinery. A "fluoroprotein" foam is being used here. This type of foam covers the burning oil quickly and does not break down easily in the heat of the fire. The fire is smothered and extinguished.

Questions

1. What is a fossil?
2. Describe the three common fossil fuels.
3. Explain why fossil fuels are really stores of solar energy.
4. What actions can we take to make our supplies of fossil fuels last longer?
5. Explain whether or not you would expect soot to burn.
6. When petrol is burnt in a car engine, the exhaust gases contain about 9% carbon dioxide, 5% carbon monoxide, 4% oxygen, 2% hydrogen, 0.2% hydrocarbons and 0.2% nitrogen oxides.
   (a) What are hydrocarbons?
   (b) The gases listed above make up only about 20% of the exhaust gases. What gas makes up most of the rest?
   (c) Name one other chemical which you might find in the exhaust gases and explain how it gets there.
   (d) Suggest reasons for the presence of
      (i) carbon dioxide
      (ii) carbon monoxide
      (iii) nitrogen oxides
   in the exhaust gases.
7. Crude oil is separated into different fractions by fractional distillation.
   (a) Name the important family of organic chemicals which make up most of crude oil.
   (b) Explain how these compounds are separated by fractional distillation.
   (c) Give the names and uses of three fractions obtained from crude oil.
8. Many countries today suffer from "acid rain".
   (a) What causes the extra acidity in the rain?
   (b) What damage can be done by acid rain?
   (c) How could this pollution be reduced?
9. Explain how you would put out the following fires:
   (a) A chip-pan fire
   (b) A waste-paper basket fire.
   (c) A fire in a stereo set.
10. Oil and natural gas may run out in your lifetime. How do you think that this would affect you?

2 Nuclear Energy

Nuclear energy is often in the news these days. The reasons for this can be seen in the two photographs. The first photograph (*figure 2.1a*) shows a nuclear power station, which is used to make electricity. Every time you switch on a light you are using electricity from a nuclear power station. The second photograph (*figure 2.1b*) shows the terrible result of using nuclear energy for fighting a war. Two nuclear bombs were dropped on cities in Japan during the Second World War. The cities were destroyed and over two hundred thousand people were killed. Even today, 40 years later, people are still dying from the effects of the bombs.

Figure 2.1a Nuclear power stations in Ayrshire, Scotland. The pylons in the foreground carry electricity to the National Grid.

Hunterston A, on the left, has been operating since 1964. Hunterston B, on the right, which consists of two advanced gas-cooled reactors, has been operating since 1977.

Figure 2.1b Hiroshima after the bomb.
About 140 000 people, nearly one third of the population, died within a year of the bomb being dropped.

Where does all this energy come from?
You will often hear "nuclear energy" called "atomic energy", and this gives the clue. In order to understand nuclear energy, it is necessary to know more about atoms.

Atoms

Everything around us is made of tiny particles called **atoms**. Atoms are made of even smaller particles called **protons, neutrons** and **electrons**. The protons and neutrons are found in the centre of the atom, which is called the **nucleus**. The electrons surround the nucleus (*figure 2.2*).
Nuclear energy is the energy that is made when the nucleus of an atom breaks up.

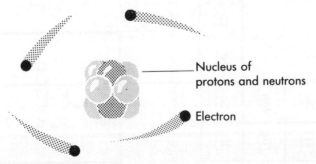

Figure 2.2 Diagram of an atom. This idea of what an atom looks like can help to explain many parts of chemistry.

I	II											III	IV	V	VI	VII	O
1 **H** Hydrogen 1																	4 **He** Helium 2
7 **Li** Lithium 3	9 **Be** Beryllium 4											11 **B** Boron 5	12 **C** Carbon 6	14 **N** Nitrogen 7	16 **O** Oxygen 8	19 **F** Fluonne 9	20 **Ne** Neon 10
23 **Na** Sodium 11	24 **Mg** Magnesium 12											27 **Al** Aluminium 13	28 **Si** Silicon 14	31 **P** Phosphorus 15	32 **S** Sulphur 16	35.5 **Cl** Chlorine 17	40 **Ar** Argon 18
39 **K** Potassium 19	40 **Ca** Calcium 20	45 **Sc** Scandium 21	48 **Ti** Titanium 22	51 **V** Vanadium 23	52 **Cr** Chromium 24	55 **Mn** Manganese 25	56 **Fe** Iron 26	59 **Co** Cobalt 27	59 **Ni** Nickel 28	63.5 **Cu** Copper 29	65 **Zn** Zinc 30	70 **Ga** Gallium 31	73 **Ge** Germanium 32	75 **As** Arsenic 33	79 **Se** Selenium 34	80 **Br** Bromine 35	84 **Kr** Krypton 36

Figure 2.3 Part of the Periodic Table of the Elements.

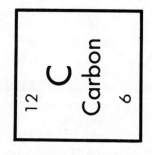

Figure 2.4 The carbon box in the Periodic Table.

Protons, neutrons and electrons

These particles have two important properties—their masses and their electric charges.

It is impossible to find out the masses of the particles directly by weighing them, because they are so small. Instead, their masses can be compared with the mass of a hydrogen atom. The mass of a hydrogen atom is approximately one **atomic mass unit**. The proton and the neutron both have nearly the same mass as a hydrogen atom. Therefore they both have a mass of approximately one atomic mass unit.

Electrons are much lighter. It takes nearly two thousand electrons to make up the mass of one proton or neutron. Electrons can be ignored when working out the mass of an atom.

Protons each have a positive electric charge. This is balanced by the electrons, which each carry a negative electric charge. Neutrons have no electric charge.

The masses and electric charges of protons, neutrons and electrons are shown in *table 2.1*.

Particle	Mass (atomic mass units)	Electric charge
Proton	1	Positive
Neutron	1	No charge
Electron	$\frac{1}{2000}$	Negative

Table 2.1   Protons, neutrons and electrons

Atoms of different elements

An atom is the smallest part of any element. Each element has its own type of atom. There are over one hundred different elements, so there must be over one hundred different atoms.

Part of the Periodic Table of the Elements is shown in *figure 2.3*.

The Periodic Table is useful because it contains information about atoms. The symbol for each atom is in a box containing two numbers as shown in *figure 2.4*. The bottom number in the box is called the **atomic number**. The atomic number is the number of protons in each atom of the element. The atomic number of carbon is six, so every carbon atom has six protons in its nucleus. If you read along the rows in the Periodic Table, you can see that the elements have been put in order using their atomic numbers.

In an atom there is an equal number of protons and electrons. This means that the atomic number is also the number of electrons in an atom.

Isotopes

Before coming to the top number in each box in the Periodic Table, it is necessary to take a closer look at the nucleus.

Atoms of the same element always have the same number of protons, but they can have different numbers of neutrons. Different atoms of the same element are called **isotopes**.

Hydrogen, the first element in the Periodic Table, has three isotopes. All of them must contain one proton (and one electron),

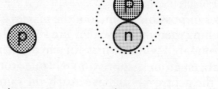

| Hydrogen-1 | Hydrogen-2 | Hydrogen-3 |
| nucleus | nucleus | nucleus |

Figure 2.5 Isotopes of hydrogen (p = proton, n = neutron).

because the atomic number of hydrogen is 1. The first isotope contains no neutrons, the second isotope contains one neutron, and the third isotope contains two neutrons (*figure 2.5*).

The main difference between these isotopes is their mass. The mass of an isotope can be calculated by adding up the numbers of protons and neutrons. Remember that protons and neutrons both have a mass of one unit, whereas electrons can be ignored.

The number of protons and neutrons together is called the **mass number**. The mass numbers of the three isotopes of hydrogen are worked out in *table 2.2*.

Isotope	*Protons*	*Neutrons*	*Mass number*
Hydrogen-1	1	0	1
Hydrogen-2	1	1	2
Hydrogen-3	1	2	3

Table 2.2 Isotopes of hydrogen

The correct way of writing the symbol for an isotope is shown in *figure 2.6*, using the isotope hydrogen-2.

mass number $\quad^{2}_{1}\text{H}$
atomic number

Figure 2.6 Symbol of the hydrogen-2 isotope.

A second example of an element with isotopes is carbon. Diagrams of three isotopes of carbon are shown in *figure 2.7*. The isotope carbon-14 is looked at later in this chapter because it is radioactive. Carbon-12 is important to scientists, because all atomic masses are measured today by comparing them with carbon-12.

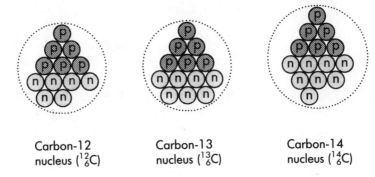

Figure 2.7 Isotopes of carbon (p = proton, n = neutron).

Relative atomic mass

Most elements are mixtures of several different isotopes. Each isotope has a different mass, because it has a different number of neutrons. The *average* mass of the atoms of any element is called the **relative atomic mass**. This is in the top number in each box of the Periodic Table. The hydrogen box is shown in *figure 2.8*.

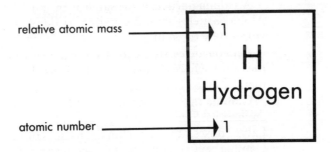

Figure 2.8 The hydrogen box in the Periodic Table.

The relative atomic mass of hydrogen is 1, even though there are isotopes with masses of 2 and 3. This is because almost all hydrogen atoms are hydrogen-1. There are so few atoms of hydrogen-2 and hydrogen-3 that they make little difference to the average mass.

Radioactivity

Sometimes the nucleus of an atom just breaks up. When this happens, energy is given out, or radiated. Atoms which break up in this way are called **radioactive**. The energy which is radiated out is called radiation.

Radiation

There are three types of radiation which can be given out by radioactive atoms. They are called alpha (α), beta (β) and gamma (γ).

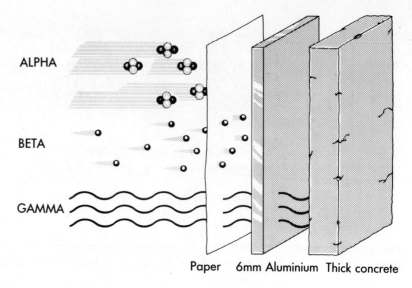

Figure 2.9 The power of alpha, beta and gamma radiation.

An **alpha particle** is the same as the nucleus of a helium atom. Alpha particles are not usually very dangerous, because they can be stopped by the skin.

Beta particles are electrons moving very fast. They can pass into the body, where they can damage cells. They can be stopped by metal about 1 cm thick.

Gamma rays are like powerful X-rays, and they can be very dangerous. Small doses to people can cause radiation sickness. Larger doses can cause skin burns, loss of hair, cancer and death. Gamma rays can only be stopped by several centimetres of lead or by thick concrete.

The power of these three kinds of radiation is shown in *figure 2.9*.

Although radioactivity can be dangerous, it can also be of great benefit to us if it is used carefully.

Isotopes in medicine

Even though radiation can cause cancer, it can be used to fight it as well, by destroying cancerous cells (*figure 2.10a*). Radioactive isotopes can also be used to give doctors information about the way that parts of the body are working (*figure 2.10b*).

Figure 2.10a Radiation therapy at the Royal Free Hospital, London. Radiation from the isotope cobalt-60 is used to treat cancer. The machine is controlled by a computer. ▶

Figure 2.10b Measuring kidney function at the Royal Infirmary, Edinburgh. The patient has been given a dose of the isotope iodine-131. The machine is used to measure the radiation from this isotope in each kidney. ▶

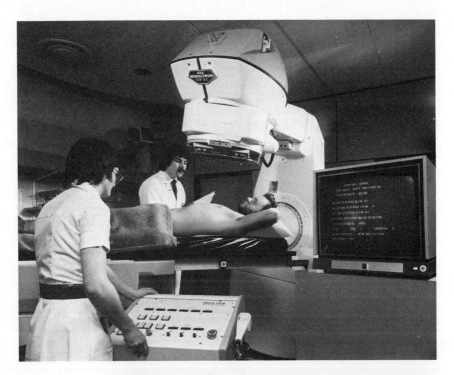
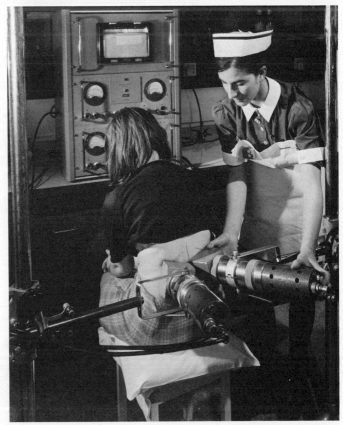

38 Chemistry in Use

Figure 2.11a Measuring pipe thickness using a portable gauge. The pipe contains radioactive caesium-137. The thicker the pipe, the less radiation gets through.

Figure 2.11b Checking the thickness of tyre cord at the Avon Rubber Company, using strontium-90.

Isotopes in industry

Many industries make use of radioactive isotopes. These uses are often based on the penetrating power of radiation. The thicker a piece of material is, the less radiation can pass through it (*figure 2.11a, b*).

Carbon dating—isotopes in archaeology

All radioactive elements get less radioactive as time goes on. This is because each time an atom gives out radiation, it changes into a different atom. When all the radioactive atoms have changed, no more radiation is given out.

Different radioactive isotopes change in this way at different speeds. The speeds are measured by a unit called the **half-life**. The half-life is the time it takes for half the radioactive atoms to change.

Half-lives can be as little as a fraction of a second, or they can be millions of years. Carbon-14 has a half-life of 5700 years. This means that, if you start out with 100 carbon-14 atoms, there will only be 50 left after 5700 years. After another 5700 years, there will be 25 left (half of 50). A graph of the change in radioactivity is shown in *figure 2.12*.

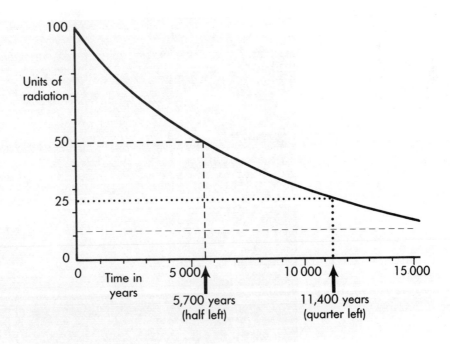

Figure 2.12 Radioactive decay of the carbon-14 isotope.

Nuclear Energy 39

40 Chemistry in Use

◀ **Figure 2.13a** Bone specimens being cut up for dating using carbon-14.

◀ **Figure 2.13b** The Round Table in Winchester, connected with the story of King Arthur. Radio-carbon dating suggested that it was made in the 13th century.

Carbon-14 can be used to find out how long ago a plant or animal died. This is called radio-carbon dating. Human bones can be dated in this way (*figure 2.13a*). Anything which was made from plants or animals can also be dated (*figure 2.13b*).

Radio-carbon dating works because all living creatures contain carbon compounds. When they die, they stop taking in any new carbon. The carbon-14 which is already there becomes less and less as time goes on, because it is radioactive. After 5700 years it is only half its original amount. The amount of carbon-14 left can be measured and used to work out how long ago the creature died.

Using Nuclear Energy
Nuclear weapons

Energy is given out when a nucleus breaks up. In an atomic bomb, energy is given out so quickly that an explosion takes place.

The first atomic bomb was made of the isotope uranium-235. Each time a uranium nucleus splits up, two or three neutrons fly out. These neutrons may then hit other uranium atoms and split them up. This is called a **chain reaction**. It goes on and on until so many atoms split at once that an explosion happens (*figure 2.14*).

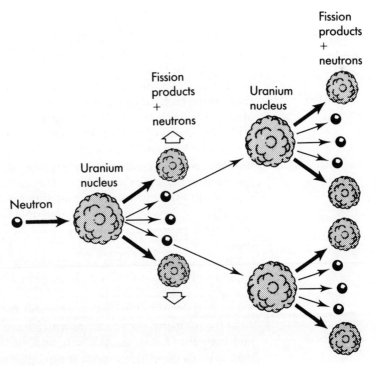

Figure 2.14 A "chain reaction".

Nuclear Energy 41

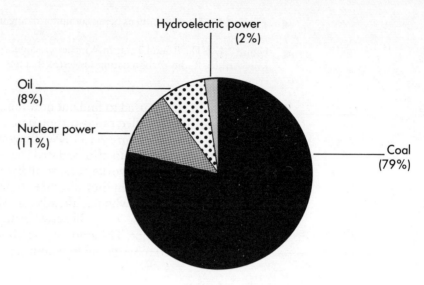

Figure 2.15 Energy supplies used to make electricity in the U.K. (1981).

Although several hundred atomic weapons have been exploded, only two have ever been used in war. One was made of uranium-235 and the other of plutonium-239. Both were dropped on Japan in August 1945. Nuclear weapons are the most powerful and terrible explosives that we know. The Hiroshima bomb had the same power as 10 000 tons of dynamite. It flattened most of the city. The world's nuclear weapons today could make over one million Hiroshima bombs. A nuclear attack by five U.S. submarines could kill about 40 million people and destroy about 60% of Soviet industry. The Russians could easily do the same damage to the U.S.A.

Atomic weapons are more dangerous than simple TNT. The radiation which they give out can affect people for years afterwards. It can even harm unborn babies. There are over 300 000 victims of the Hiroshima bomb still living.

Nuclear weapons have such horrible effects that most people want to see them destroyed. Some people think that all the weapons in the U.K. should be destroyed immediately, even if no other country does this. The Campaign for Nuclear Disarmament (CND) is the best known group of people who think like this. Up to 100 000 people at a time have attended CND meetings in recent years.

Other people think that we should only abolish our nuclear weapons if other countries do so as well. These people say that a major war in Europe has been avoided for about 40 years only because of the possible effects of nuclear weapons.

At the moment, the countries which have nuclear weapons, including the U.S.A., the U.S.S.R., and the U.K., cannot agree to stop making them. They also cannot agree to get rid of the weapons which they already have.

Nuclear power stations

In a nuclear power station, the energy which is released when atoms split up is used to make electricity. Nuclear power provides over 10% of the electricity in the U.K., although most of the electricity is made by burning coal (*figure 2.15*).

The first large nuclear power station was opened at Calder Hall in Cumbria in 1956. About 20 have been built in the U.K. since then (*figure 2.16*).

Figure 2.16 Nuclear power stations in the U.K.

Nuclear Energy 43

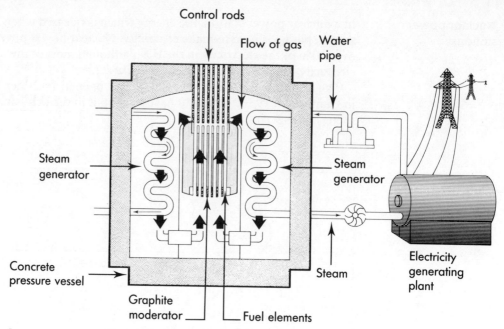

Figure 2.17 An Advanced Gas-cooled Reactor (AGR).
The gas is heated up as it flows through the reactor (thick arrows in the diagram). The gas passes its heat to water, which is turned into steam (steam generator in the diagram). The steam turns a turbine to make electricity (electricity generating plant in the diagram).

Almost all nuclear power stations use uranium as their fuel. The uranium is made into rods called fuel rods. These fuel rods are placed in the reactor of the power station. Inside the reactor the uranium atoms split up and give out heat energy.

The reaction is maintained by using graphite, which is put into the reactor with the fuel rods. The graphite is used to make sure that the neutrons, which are released when one nucleus splits up, can hit another nucleus to keep the reaction going. Graphite is called the *moderator* for this reaction.

The whole reaction can be controlled by using rods of boron. When the rods are pushed into the reactor, they absorb the neutrons and shut down the reactor (*figure 2.17*).

In many power stations, the heat energy in the reactor is used to heat up carbon dioxide. The hot gas drives a steam generator, and the steam produced is used to make electricity.

This sort of power station can only use the isotope uranium-235, which is rare. Less than 1% of uranium is uranium-235. The rest is uranium-238. The world supply of uranium is small. We probably only have enough for 50 years in ordinary nuclear power stations. Because of this, scientists have made a new type of nuclear power station, the fast-breeder, which can use uranium-238. Using the fast-breeder, our uranium could last for about 2000 years. This sounds fine, but there are two problems. Is nuclear power reasonably safe, and is it necessary?

Is nuclear power safe?

Many people are worried that there are too many risks from nuclear power stations, especially from fast-breeders. They say that we should not build any more. Although it is not possible for a nuclear power station to explode like an atomic bomb, there are other dangers.

The first danger is the connection between power stations and atomic bombs. In an ordinary power station, and in the special factories used to make the fuel for the fast-breeder, the isotope plutonium-239 is made. Plutonium-239 can be used to make atomic bombs. Any country with a nuclear power station is half-way towards making atomic bombs. In addition, these places could be targets for terrorist groups.

The second danger is the leakage of radioactivity from a nuclear power station. Leaks have already happened. Some people say they are serious. Others say that there is nothing to worry about.

The third danger is the radioactive waste from these power stations. The waste will have to be stored safely for hundreds of years. It cannot just be dumped on a rubbish tip like ordinary waste, because it is too dangerous. People are finding better ways to store it, but they cannot stop it from being radioactive. Is it fair of us to leave this dangerous waste for others to look after, years after we are dead?

Is nuclear power necessary?

There are risks in using nuclear power. However, there are also risks in not using it. We will need something to replace energy from natural gas and oil as they start to run out. Coal could help us for some time, but there could be serious pollution problems with coal (acid rain, p.23, and greenhouse effect, p.22).

Try to imagine what would happen to our society if we ran short of energy. There could be serious riots, or worse, as people fought for the energy that was left. It is difficult to know which risk is the greater.

Some countries are already deciding not to use nuclear power. For example, the Austrians have voted not to build any nuclear power stations. How do you think that people in this country would vote? What would be their reasons? You could organize a survey to find out.

Questions

1   Complete the following table about atomic structure:

	$^{1}_{1}H$	$^{14}_{6}C$	$^{24}_{12}Mg$	$^{31}_{15}P$	$^{235}_{92}U$
Number of protons					
Number of neutrons					

Nuclear Energy

2. The element chlorine exists as two isotopes.
   (a) Complete the following table about these two different kinds of atoms:

	Atomic number	Mass number	Number of protons	Number of neutrons
Isotope 1	17	35		
Isotope 2			17	20

   (b) Chlorine has a relative atomic mass of 35.5. Explain this by referring to the mass numbers in the above table.

3. Name the three types of radiation which are produced by radioactive atoms. Explain what each type of radiation contains.

4. If you were working with γ-rays, what safety measures would you take?

5. Explain how radiation could be used to control the thickness of a paper sheet produced by a machine.

6. A chemist starts with 12 g of a radioactive isotope which has a half-life of 10 days. How much of it is left after 30 days?

7. The following measurements were made on a radioactive element:

Hours from start of experiment	Radioactivity
0	100%
1	79%
2	63%
4	40%
8	16%
10	10%
15	3%

   Plot a graph of these results. Use the graph to work out the half-life of this isotope.

8. Imagine that there are plans to build a nuclear power station near your home, which is in an area of high unemployment. List the advantages and disadvantages of building this power station. Using your list, explain whether you would be in favour of the plan or against it.

3 Alternative Energy Supplies

The Energy Crisis

Few people expect the world's supplies of natural gas and oil to last much more than 50 years, and yet about 65% of the world's present energy supplies come from these two fuels (*figure 3.1*).

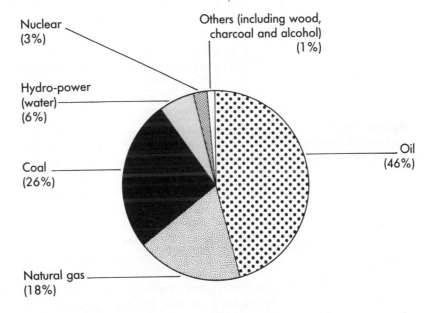

Figure 3.1 The world's use of energy supplies (1980).

Coal could be available for a little longer, perhaps three hundred years. This may seem a long time to you, but remember that people have been living on this planet for millions of years already.

There is another problem with fossil fuels. If we continue to burn them so quickly, many scientists believe that we could overheat our atmosphere. The world's weather could be seriously affected (greenhouse effect, p.22).

An obvious alternative to fossil fuels is to use uranium in nuclear power stations (p.43). There are problems here as well. The supply of uranium for ordinary power stations will probably run out before our oil. We could make the uranium last for perhaps

2000 years by using fast-breeder reactors, but people are worried by the radioactive waste and the connection with nuclear weapons.

We are facing an energy crisis. There are two things we can do about it. One is to use energy as sparingly as possible. The second is to look for other supplies of energy. This chapter is about new energy supplies, but conservation is also important. A recent American report says that a third of all energy used in the U.S.A. could be saved by better design of new buildings and alterations to older ones.

What makes a good energy supply?

There are three important points to think about:

1. The energy supply should be replaceable.
   This means that it can always be renewed and will not run out. Fossil fuels and uranium are not replaceable.
2. There should be plenty of the energy supply and it should be reasonably cheap to use.
3. Using the energy should cause as little pollution or damage to the environment as possible.

Replaceable Energy Supplies
Wood and charcoal

Wood, and charcoal which is made from it, provide only a small part of the world's energy (*figure 3.1*). This does not mean that they are unimportant. They are the two most important fuels for many less-developed countries. In places like Thailand, Tanzania and Kampuchea they provide 90% of the fuel supply. They are vital for cooking. Over half the energy used for cooking in the whole of India comes from firewood.

Wood and charcoal are renewable fuels, because trees can grow again. They seem to be a good choice. However, in many countries trees are cut down faster than they can grow. Gambia has already lost 96% of its forests. In the Sudan, wood is so scarce that people gather it from 100 km away (*figure 3.2*).

Wood is not important only to the less-developed countries. About half the wood cut down each year is burnt. Most of the rest is turned into pulp to make paper for the developed countries. Much of this is thrown away and wasted, although people are starting to realize that wasting paper means wasting trees.

If they are used carefully, wood and charcoal can be valuable replaceable fuels. They are not being used carefully enough at present. We are destroying a piece of tropical forest the size of the Netherlands each year. Apart from giving us less wood, this could also badly affect the world's climate.

Methane—making our own natural gas

The methane which comes out of rocks as natural gas cannot be replaced, but there are other ways of obtaining this gas.

If you have looked closely at a marsh or pond, you may have noticed bubbles of gas rising to the surface. These are bubbles of methane, produced when plants and animals rot. This is how our supplies of natural gas were made, millions of years ago.

Figure 3.2 A camel train carrying wood in the Sudan.

Figure 3.3 A van being filled up with methane from a gas pump. The equivalent of a gallon of petrol costs as little as £1. A cylinder of gas lasts for about 80 km, enough for local journeys. If it runs out, the engine can be changed to run on petrol at the flick of a switch.

Large amounts of methane can be made if plant or animal remains are collected and allowed to rot. In the U.K. and in many other countries, methane is made in sewage works. Some sewage works run all their machines and fuel their vans with methane (*figure 3.3*).

Alternative Energy Supplies 49

Figure 3.4a Chinese technicians making a biogas digestor. They are standing in the part which will become the fermentation chamber.

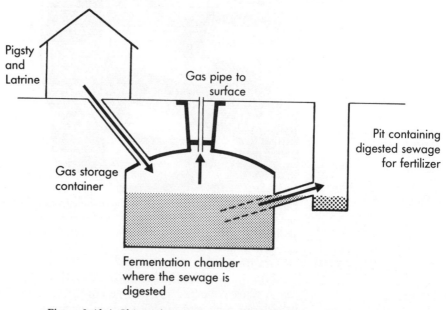

Figure 3.4b A Chinese biogas digestor. Waste from animals and humans is fed into the digestor. It is fermented to produce the gas (about 60% methane). After fermentation, the waste is returned to the land as fertilizer.

Methane is the best of the hydrocarbon fuels, because it gives out the least pollution when it burns.

In countries like India and China, cow dung and human waste are allowed to rot in small tanks called gas digestors (*figure 3.4a, b*). The methane can be used for cooking and for lighting. The rest of the waste can be used as a fertilizer. Small gas digestors like these could provide cheap fuel and fertilizer for many communities.

Alcohol—fuel from food

Alcohol can be made from sugar by a method called **fermentation** (p.191), as every home-brewer knows. People have used this reaction for thousands of years to make drinks like beer and wine.

In Brazil today, alcohol is not only made for drinking. Alcohol burns well, with a clear flame, and the Brazilians use it as a fuel. They do this because they can produce more sugar cane than they need to eat, and because it is expensive to buy oil from abroad.

Car engines have to be altered to run on pure alcohol or on a mixture of alcohol and petrol. The alteration is quite simple. By the end of this century, alcohol should provide half the energy used for transport in Brazil (*figure 3.5*).

Figure 3.5 Alcohol-powered transport in Brazil.

Alternative Energy Supplies 51

Making alcohol from sugar is useful for Brazil, but it cannot solve the world's energy problem. It is only possible to do it in countries like Brazil, which have plenty of land for growing food.

If all the farm land in the U.K. were used for this purpose, it would supply only about a tenth of the energy needed. It makes more sense to grow food in the U.K.

Hydrogen—fuel of the future?

Hydrogen is an ideal fuel in many ways. Only water is produced when it burns, so there is no pollution. There is plenty of hydrogen on this planet, but it is combined with oxygen in sea water. It could be made from sea water, using electricity, and we would not run out of it. Unfortunately, this is too expensive at the moment, because electricity has to be made first. Scientists are trying to find cheaper ways, using energy from the sun, to make the electricity.

There are two problems with hydrogen as a fuel. It is difficult to store, because it is a gas, and it can explode quite easily. If scientists can find ways round these problems, hydrogen could be the fuel of the future.

Energy without Chemistry

All the energy supplies mentioned so far are chemicals. If energy can be made without burning chemicals, then we can save those chemicals for other uses. The energy supplies described so far, except for uranium, are also all stores of solar energy in some way. Chemicals are not the only possible stores of solar energy.

Water, tides, wind and waves

Energy from the sun turns water into clouds, which fall as rain into lakes and rivers. In mountain areas, running water can be used to drive turbines which make electricity. This is called hydroelectric power. It is important in mountainous countries like Sweden, which gets 25% of its energy in this way. There are hydroelectric power stations in the mountains of the U.K. (*figure 3.6*), although they provide less than 10% of the energy used in the U.K.

The sun also causes tides and winds, which in turn cause waves. All these can be used to make electricity. They could not supply the world's energy needs alone, but they could help. Towns and villages near to any of these supplies of energy could have their own small stations. This would probably be better than building large power schemes. An enormous barrage across the River Severn to catch the tide could make about 7% of the electricity for the U.K. However, it would be expensive to build and it would damage the animal and plant life in the Severn estuary.

Geothermal energy

The rocks hundreds or thousands of metres below the earth's surface are hot. If cold water is pumped down, it comes up again as hot water or even steam. Experiments have started in Cornwall to see if it is worth doing in the U.K. (*figure 3.7*).

Figure 3.6 The Luichart dam, part of the hydroelectric scheme in the north of Scotland. The water which is held by this dam is allowed to flow downhill, driving a turbine to make electricity.

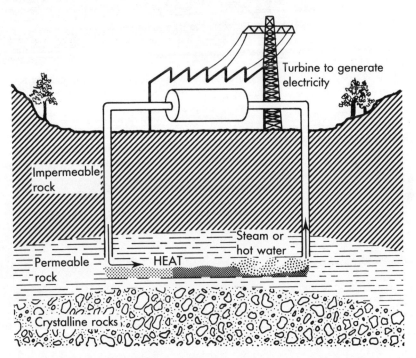

Figure 3.7 Geothermal energy. Cold water is forced down a pipe into a permeable rock like limestone or sandstone. Heat from the rocks warms up the water until it passes into a pipe carrying it to the surface. Hotter, impermeable rocks like granite can also be used. The rocks are fractured using explosives, so that water is channelled through them.

The hot water or steam can be used for heating or for generating electricity. Many apartment blocks and offices in Paris are heated by geothermal power. Much of the electricity for Italian Railways is generated in this way.

Alternative Energy Supplies 53

Solar energy

In the end, everything comes back to solar energy—the power of the sun. There is plenty of it reaching us, far more than we need, but it is not always easy to trap cheaply. Solar energy for heating or cooling single houses or small communities is probably the best use at present (*figure 3.8a, b*). Solar heating is cheaper than electricity in most parts of the world—and it does not use fossil fuels.

Figure 3.8a Solar panels at Albuquerque, New Mexico. Even in many less sunny parts of the U.S.A., enough solar energy falls on buildings to heat them fully.

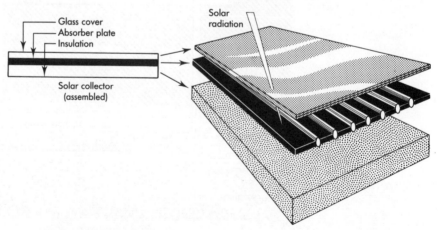

Figure 3.8b Design of a solar collector used for water heating. Glass covers allow heat through but do not allow it to pass back. Greenhouses work in the same way. The absorber plate is made of copper or aluminium, two metals which absorb heat well. The plate is coated black to absorb the maximum amount of heat. Water flows through copper pipes in the absorber and takes in the heat. The hot water can then be used as required. Underneath the absorber plate is a layer of insulation. This stops heat escaping through the base and sides of the collector.

PART B
The Chemistry of Materials

Once we have our supplies of energy, we can start to shape the materials around us and make the things we need. There are hundreds of materials for us to choose from, including metals, stone, glass, salt, plastics and rubber, some of which are natural and some man-made.

Metals can be made into objects like tools, wires, trains and ships. Ordinary salt can be used to make bleaches, soaps and many other chemicals. In recent years, crude oil has given us a whole new range of materials. We use plastics, rubbers and man-made fibres to make packaging, tyres, clothes and thousands of other everyday objects. This section starts with this modern group of materials, all made from oil.

A very unusual use of material—the artist Christo's covering of cliffs with canvas.

4 Chemicals from Oil

Look round your house today and you will find dozens of objects made out of plastic. These could include telephones, floor tiles, carpets, radio and TV cases, bowls, bottles, shelves, chairs and table tops, to name just a few. Look at your clothes and you will find man-made fibres like nylon, terylene and acrylics. Look over a car and you will find about 50 kg of plastics, rubber on the tyres, and, if it is winter, antifreeze in the radiator.

All these things have been made from oil. We burn about 90% of our oil as a fuel. The other 10% goes to make plastics, man-made fibres, rubber, paints, glues, pesticides, solvents, detergents and many other chemicals. It is difficult to imagine life without them (*figure 4.1a, b*).

Figure 4.1a How many things in this picture could have been made from the chemicals in crude oil?

Figure 4.1b The oil refinery at Berre L'Etang in France. The crude oil, or petroleum, is separated into different fractions. Most of these are burnt as fuels, while others go to the chemical industry. Chemicals which are made at the refinery itself include rubber and detergents. Making chemicals from oil is often called the "petrochemical" industry.

The Chemicals in Crude Oil

Crude oil is a mixture of chemicals called hydrocarbons. Most of these belong to a family of chemicals called alkanes, which contain chains of carbon atoms (p.11).

Carbon is an unusual element, because so many different chemicals can be made from it. This is because carbon atoms are the only atoms which can easily form into long chains or rings. There are several million different chemicals known to us today and most of these contain carbon. At least 15 000 of them are made and sold by companies for our use. Chemicals like these, containing carbon, are called **organic chemicals**. This name is used because living organisms are based on compounds of carbon. Oil is rich in carbon, because it was formed from decaying creatures (p.2).

All the objects mentioned at the beginning of this chapter are made from organic chemicals, and 90% of the world's organic chemicals are made from oil. In many ways, oil is too precious to burn.

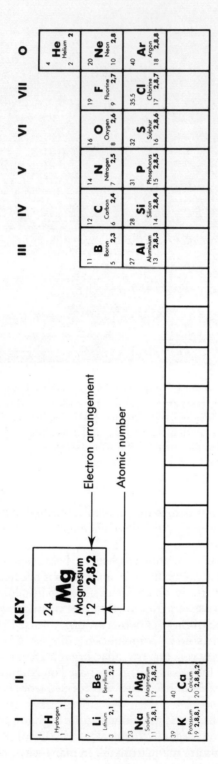

Figure 4.2 The Periodic Table, showing the arrangement of electrons in atoms.

58 Chemistry in Use

A closer look at hydrocarbons

All hydrocarbons are found as molecules, which are simply groups of atoms joined together. In the case of methane CH_4, the simplest hydrocarbon, each molecule is made of one carbon atom joined to four hydrogen atoms.

In order to understand how the atoms are held together, it is necessary to look carefully at the electrons in the atoms. It is electrons which are involved in holding atoms together in molecules.

Building Molecules from Atoms

Electrons in atoms

The atomic number of an element tells you how many electrons there are in every atom of that element (p.33). Hydrogen is the first element in the Periodic Table, so it has an atomic number of 1. This means that every atom of hydrogen has just one electron. Carbon has an atomic number of 6, so every carbon atom has six electrons.

The number of electrons in an atom is important, but it is even more important to know how these electrons are arranged in the atom.

Arrangement of electrons in atoms

Most atoms contain many electrons. It helps to imagine that these electrons are arranged in layers round the centre of the atom (the nucleus). The different layers of electrons are called **electron shells**. While you are thinking about electron shells, it is useful to look at the Periodic Table. A section of it, giving the first 20 elements, is shown in *figure 4.2*.

1   The first electron shell, which is nearest to the nucleus, can hold up to 2 electrons. The single electron in an atom of hydrogen (H, atomic number 1) is in this shell. Both electrons in an atom of helium (He, atomic number 2) are in the first shell. Helium atoms therefore have a first shell of electrons which is full.

2   The second electron shell is used if there are more than two electrons in an atom. This shell can hold up to 8 electrons. Atoms of the element lithium (Li, atomic number 3) contain three electrons. Two of them are in the first shell and the third one starts the second shell. Notice that lithium also starts a new row in the Periodic Table. Whenever a new row in the Periodic Table is started, a new shell of electrons is started as well.

The arrangement of electrons in lithium atoms can be written 2,1. This shows that there are two electrons in the first shell and one in the second shell. Using these ideas, the electrons in beryllium atoms (Be, atomic number 4) are arranged 2,2. In boron atoms (B, atomic number 5), the electrons are arranged 2,3. Atoms of neon (Ne, atomic number 10) have a full second shell of electrons. Neon has the electron arrangement 2,8.

3   A third shell is needed for atoms of sodium (Na, atomic number 11). Sodium has the electron arrangement 2,8,1. Notice that sodium starts a new row in the Periodic Table. This row ends with argon (Ar, atomic number 18), which has the electron arrangement 2,8,8.

4   The fourth shell and the fourth row of the Periodic Table are started with potassium (K, atomic number 19).

The electron arrangements in the atoms of the first 20 elements are shown in *figure 4.2*.

Another way of showing how electrons are arranged in atoms is to use the diagrams shown in *figure 4.3*. In these diagrams a circle represents a shell of electrons. Each electron in the atom is shown by a small filled circle.

Joining atoms together

The elements helium, neon and argon give a clue about how atoms join together. Atoms of helium, neon and argon have full electron shells and they hardly react with anything. It seems that atoms with full shells of electrons are stable and, therefore, unreactive. We can use this idea to explain why certain atoms join together.

When elements combine together, their atoms usually end up with full shells of electrons. Some atoms can gain or lose small numbers of electrons, generally only one or two, in order to get full shells. Other atoms, including carbon, cannot gain or lose enough electrons. Carbon atoms would need to gain or lose four electrons to get full shells, and this is not possible (p.107). Instead, carbon atoms share electrons with other atoms. The electrons are shared in pairs, one from one atom and one from the other. The shared pair of electrons holds the atoms together. This is called a **covalent bond**.

Covalent bonds

A molecule of methane is held together by four covalent bonds (*figure 4.4*). Methane is an example of a **covalent compound**.

Each electron in the second shell of the carbon atom pairs with the electron from a hydrogen atom. This gives four pairs of electrons holding the atoms together, making up four covalent bonds. You can see from *figure 4.4* that each hydrogen atom has a share in two electrons. Its shell is therefore full, like an unreactive helium atom. In the same way, each carbon atom has a share in eight electrons in its outer shell. This shell is also full, like an unreactive neon atom.

Carbon atoms always need four electrons to get a full outer shell, so they always form four covalent bonds. Hydrogen always forms one covalent bond, because it needs only one electron to get a full shell. The number of bonds formed by an atom is sometimes called the **valency** of that element.

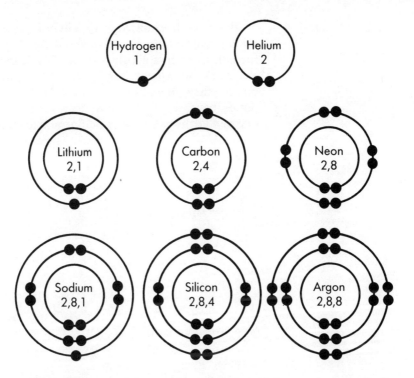

Figure 4.3 One way of showing the arrangement of electrons in atoms. Each circle represents one shell of electrons. Electrons are shown by the small filled circles.

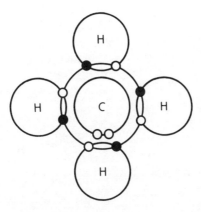

Figure 4.4 Covalent bonding in a molecule of methane (CH_4). Electrons from the carbon atom are shown by small open circles. Electrons from the hydrogen atoms are shown by small filled circles.

Chemicals from Oil 61

Small molecules

Methane is one example of a small molecule. Diagrams of some other simple molecules are shown in *figure 4.5*. The structural formula of each molecule is shown beside the diagram of the electron arrangement. In a structural formula, each line represents one covalent bond. Notice that nitrogen forms three covalent bonds, because it needs three electrons to complete its outer shell. Oxygen forms two bonds and fluorine forms one bond.

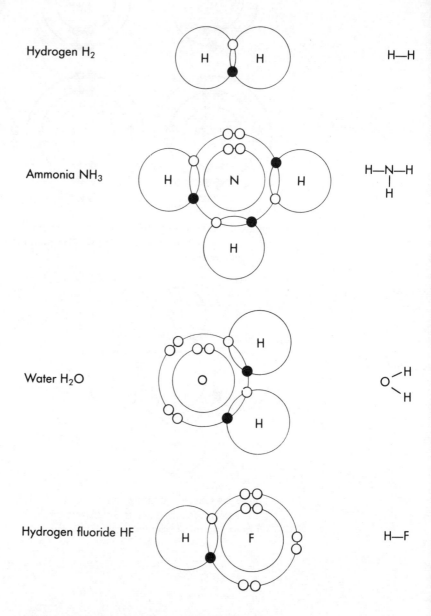

Figure 4.5 Covalent bonding in some simple molecules. Notice how every atom in these molecules has got full shells of electrons.

62 Chemistry in Use

Building larger molecules

The alkanes which are found in crude oil are larger molecules than methane. The structural formula of any alkane can be built up by using the information that carbon atoms form four bonds and hydrogen atoms form one bond (*figure 4.6a*).

The chains of carbon atoms do not need to be straight, provided that each carbon atom forms four covalent bonds. This means that different molecules can be made from exactly the same atoms. Different molecules containing the same atoms are called **isomers** (*figure 4.6b*).

ALKANE	MOLECULAR STRUCTURE	STRUCTURAL FORMULA
Ethane	C_2H_6	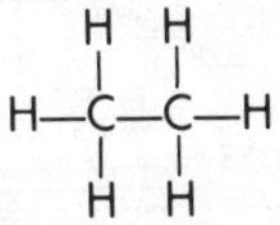 H-C-C-H (with H's above and below each C)
Propane	C_3H_8	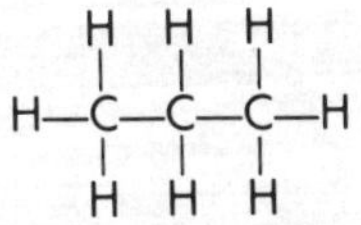 H-C-C-C-H (with H's above and below each C)
Butane	C_4H_{10}	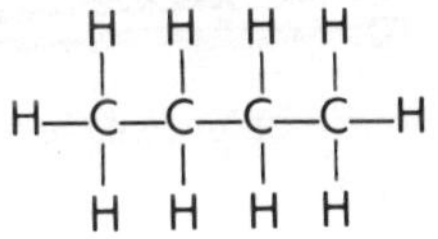 H-C-C-C-C-H (with H's above and below each C)
Octane	C_8H_{18}	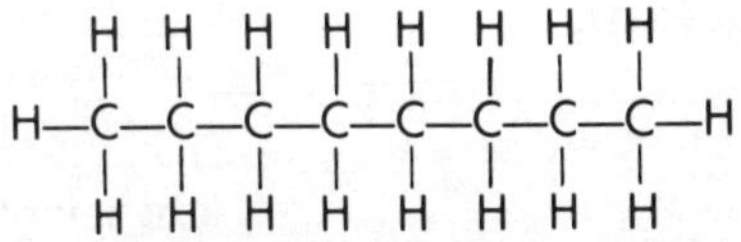 H-C-C-C-C-C-C-C-C-H (with H's above and below each C)

Figure 4.6a Alkanes. Notice that every carbon atom has four covalent bonds and every hydrogen atom has one covalent bond.

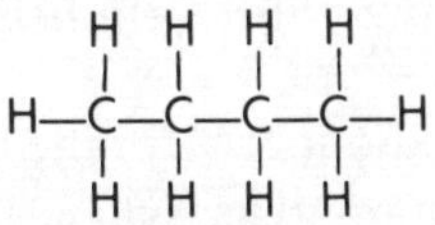

Butane (C_4H_{10}) — straight chain

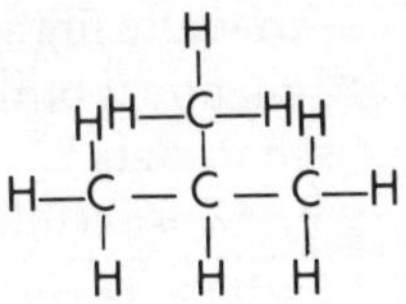

Isomer of butane (also C_4H_{10}) — branched chain

Figure 4.6b Isomers. Isomers have the same molecular formula, but the atoms are arranged differently. Notice that every carbon atom still has four covalent bonds and every hydrogen atom has one covalent bond. The difference is that one isomer has a branch in its chain of carbon atoms. Hydrocarbons with branched chains are important in petrol. They burn more smoothly than hydrocarbons with straight chains.

Figure 4.7a The catalytic cracking unit at the Clyde refinery, Australia.

Cracking

Alkanes from crude oil are not much use directly, except for burning. They do not react easily, so they are difficult to turn into other, more useful, chemicals.

Oil scientists have found a way round this problem. It is called cracking.

Naphtha—the "feedstock" for cracking

Naphtha is one of the fractions which is made by refining crude oil (p.13). It is one of the easiest fractions to crack, and it is the most commonly used fraction in Europe. Part of the naphtha is cracked to make high-grade petrol (p.14). The rest is cracked to make the chemical building blocks of plastics, fibres, rubbers and other materials.

Most crude oil contains only about 20% of the naphtha fraction. Other fractions are now being cracked as well, as the demand for petrol and chemicals rises.

The cracker

Naphtha is a mixture of many different hydrocarbons. Most of them are molecules which have chains of between four and twelve carbon atoms. These molecules are broken down or "cracked" into smaller, more useful, pieces in the cracker.

All that is needed to do this is heat. In one method, called steam cracking, a mixture of naphtha and steam is passed through red

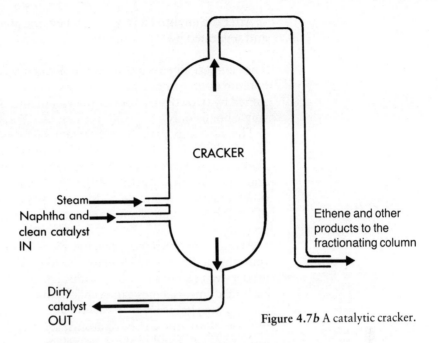

Figure 4.7b A catalytic cracker.

hot pipes at about 800°C. The other method is called catalytic cracking or "cat-cracking". **Catalysts** are chemicals which speed up chemical reactions without being used up themselves. By using a catalyst, it is possible to carry out the reaction at a lower temperature. This saves energy.

Cracking the alkanes in naphtha makes a completely new family of organic compounds, as you can see in *figure 4.8*. These new compounds contain two covalent bonds between some carbon atoms instead of just one. They belong to a family of organic chemicals called the **alkenes**.

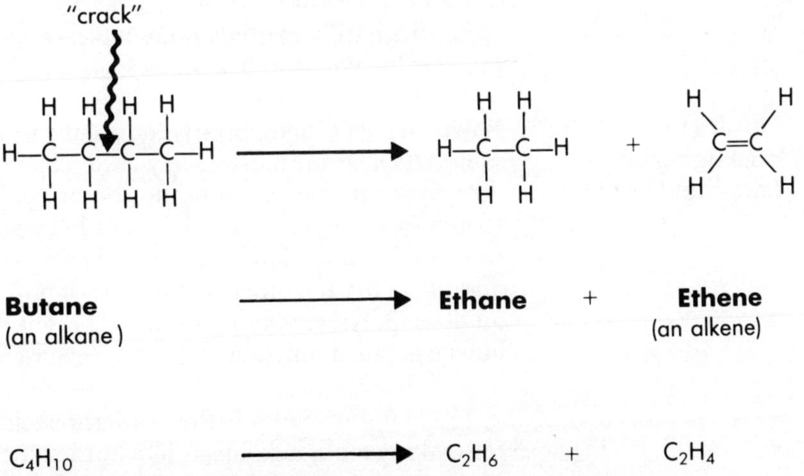

Butane → **Ethane** + **Ethene**
(an alkane)   (an alkene)

C_4H_{10} → C_2H_6 + C_2H_4

Figure 4.8 A typical reaction in a cracker.

Chemicals from Oil

Alkenes

After the naphtha has been cracked, the alkenes are collected and separated by fractionation.

The simplest alkenes are shown in *figure 4.9*. The molecules all contain double bonds, which means that some carbon atoms are joined by two covalent bonds instead of one. Alkanes contain single bonds only.

The most important of these alkenes is ethene C_2H_4, because it can be turned into polythene, PVC, polystyrene, antifreeze and alcohol, amongst other chemicals. Ethene is the main product of cracking naphtha, together with smaller amounts of propene and other alkenes. It is strange to think that much of our modern way of life is built round ethene, a sweet-smelling gas which is made by cracking oil.

Alkenes are much more reactive than alkanes, because they contain double bonds. The extra bond in the double bond can be used to join alkenes to other atoms, instead of joining the two carbon atoms. Alkenes can even join together in this way. When they do this, long chains of carbon atoms are formed. Molecules like this, with long chains of atoms, are called **polymers**. All our modern plastics and rubbers, as well as many paints and glues, are polymers.

Plastics

It is not only the obvious things like squeezy bottles, washing-up bowls, telephones and carrier bags which are made of plastics. Man-made fibres like terylene and nylon are also plastics. Plastics are man-made materials which can be turned into any desired shape while they are in an almost liquid or "plastic" form. They can be made into thin films for packaging, moulded into shapes like cups, or spun into fibres to make clothes, carpets and ropes. Some examples are shown in *figure 4.10*.

Plastics hardly existed about 40 years ago. It is difficult to imagine life without them now.

Plastics for packaging and household goods

Polythene, PVC and polystyrene are the three best known plastics.

Polythene is the most widely used plastic today. It can be made directly from ethene. Thousands of molecules of ethene are made to join together into a long chain called a polymer. This reaction is called **polymerization**. The ethene molecules polymerize by simply adding on to each other, so this sort of polymerization is called **addition polymerization** (*figure 4.11*). The polymer made from ethene is called poly(ethene), although it is more commonly known as polythene.

The major uses of polythene are for packaging (poly-bags) and for making moulded objects like bottles and cups. It also has some more unusual uses. Surgery to replace heart valves and other body parts could not be done without polythene (p.216).

ALKENE	MOLECULAR FORMULA	STRUCTURAL FORMULA
Ethene	C_2H_4	$\begin{array}{c}HH\\ C=C\\ HH\end{array}$
Propene	C_3H_6	H₂C=CH–CH₃
Butene	C_4H_8	H₂C=CH–CH₂–CH₃

Figure 4.9 Alkenes.

Figure 4.10 Some useful objects made from different plastics.

Chemicals from Oil 67

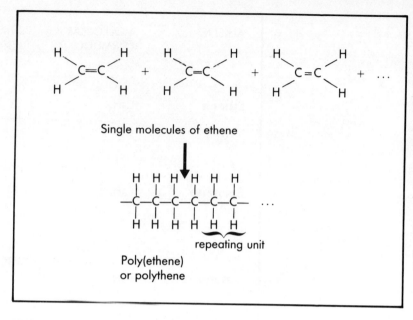

Figure 4.11 Polymerization of ethene to polythene.

PVC, which stands for *polyvinyl chloride*, is another much-used plastic. PVC, like polythene, can be made from ethene, although chlorine is needed as well. Making PVC is one of the most important uses of chlorine (p.132).

PVC is made by polymerizing the chemical chloroethene, which is itself made from ethene. This is another addition polymerization, just like the making of polythene (*figure 4.12*). The proper name for this polymer is poly(chloroethene), although it is usually called PVC.

The main uses of PVC are for making floor tiles, pipes and waterproof material or imitation leather for clothes, shoes and bags. It is also used as an insulator round electrical wires. The computer industry could not have grown so fast without plastics as insulators.

Polystyrene is the last of the three major plastics. It is also made from ethene. Polystyrene cups for hot drinks and light polystyrene foam for packaging and house insulation are common sights today.

Heat-resistant plastics

Most forms of polythene, PVC and polystyrene are not resistant to heat. If they are heated carefully, they will melt and can be reshaped. Polymers like this are called **thermoplastic polymers**. Thermoplastic polymers can also be burnt easily. If this sort of polymer is used in furnishings, it can be a fire risk. The materials burn quickly and they may produce poisonous fumes.

There are other polymers, called **thermosetting polymers**, which are resistant to heat. Thermosetting polymers cannot be reshaped

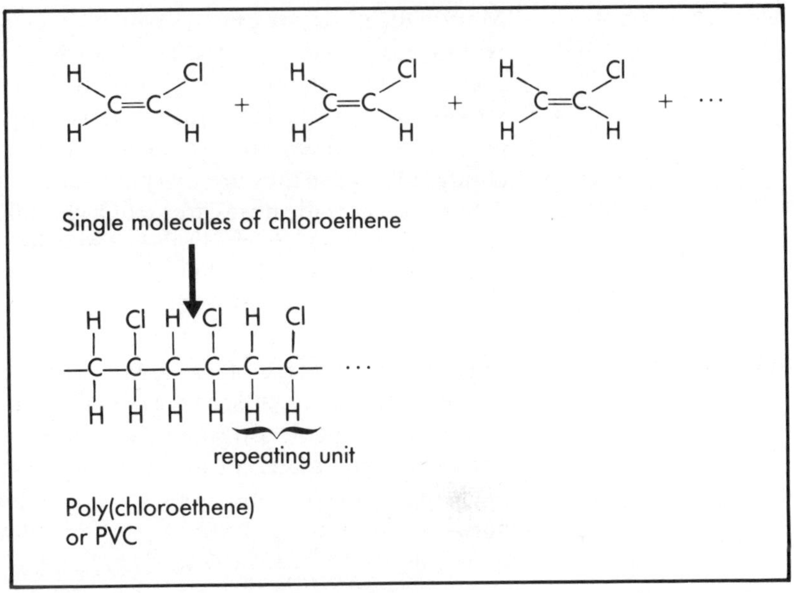

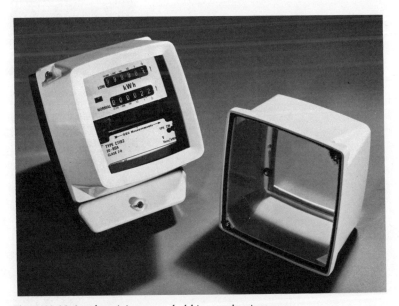

Figure 4.12 Polymerization of chloroethene to PVC.

Figure 4.13 An electricity meter held in a melamine case.

by heating them after they have been made. Polymers of this sort, including melamine, can be used for making things like table tops and electrical fittings (*figure 4.13*).

All polymers, including thermosetting polymers, will eventually burn, because they all contain carbon. If you have looked inside an electric plug which has blown a fuse, you may have seen a blackened mark where the plastic has been scorched.

Chemicals from Oil 69

Man-made fibres

Man-made fibres are plastics which have been spun into threads. Nylon and Terylene are the best-known examples of these polymers. Clothes made from these fibres can be shrink-proof, crease-proof and easy to wash, but they are not usually as warm as natural fibres. They are also more of a fire risk, especially for children, because they are thermoplastic polymers.

Some man-made fibres are strong enough to weave into sacks, carpets and ropes. Rock climbers today use nylon ropes which are much stronger than the old hemp ropes (*figure 4.14*).

A plastics crisis?

Our present standard of living would not be possible without plastics. Plastics are able to replace natural materials including wood, metal, wool, cotton, leather and stone. There is not enough of these natural materials to satisfy the demand for clothing, packaging and consumer goods today, so plastics are needed. It hardly makes sense to burn 90% of the oil which gives us these materials.

Figure 4.14 The author putting his faith in nylon ropes on a climb called "The Mole" in North Wales.

When the oil runs out, other ways of making plastics will be needed. A likely possibility is a return to coal. Many of these materials were made from coal before oil became cheaper (p.21). A more exciting idea is the use of bacteria to make chemicals which can be turned into plastics. This is part of a whole area of science called *biotechnology*—using living organisms to make the chemicals we want.

Plastic waste and land pollution

Most plastics are long-lasting. They do not rot away if they are thrown down as litter. They are not **biodegradable**, which means that they cannot be broken down by creatures like bacteria in the soil. For this reason, plastic waste should always be thrown into a bin, so that it can be removed with other rubbish.

Plastic waste forms about 5% of ordinary household rubbish. It is usually just put directly onto a rubbish dump. There are two reasons for this. One is that it is difficult to separate from other rubbish. The other is that it is difficult to recycle, or use again, even if it has been separated. A good use of plastic waste is to burn it for energy (*figure 4.15*).

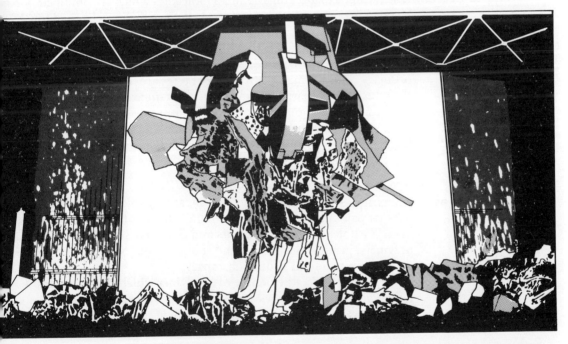

YOU COULD CUT YOUR COMPANY'S FUEL BILLS BY TALKING RUBBISH.

Over the last few years a number of companies have found that by using rubbish as fuel they can save on energy costs and become more competitive.

Of course, the manner in which individual companies went about it varied.

But as either an energy manager or a production manager it could pay you to look at the many examples of where waste has been used as a fuel in the Energy Conservation Demonstration Projects Scheme (ECDPS).

For example, at Ford's Dagenham plant and other plants in the area, they produce more than 300 tons of combustible factory waste every week. Consisting in the main of paper, cardboard, wood, plastics and rubber.

As an alternative to being tipped at a local landfill site, this material is fed into special incineration equipment. Which, in addition to various other processes, should cut their annual fuel costs by some £400,000.

The ECDPS includes many other projects that show how both solid and liquid waste can be used as a fuel.

Send the coupon for details of the many schemes to help your company make better use of energy and become more competitive.

To: Dept. of Energy, London SW20 8SZ.
Please send me more information on how I can make better use of energy.
Name _____
Job Title _____
Address _____
Tel. No. _____ **ENERGY**

Figure 4.15 Energy can be saved by burning rubbish containing paper, cardboard, wood, plastics or rubber.

Chemicals from Oil

Plastic waste is only a small part of the 56 million tonnes of general rubbish which we make each year. 90% of this rubbish is used for "landfill". In other words, it is put on rubbish dumps. New sites for dumping rubbish are becoming harder and harder to find. It would make more sense if we could recycle more of our waste. It is often expensive to do this, but much more could be done.

Paper forms 60% of the volume of household rubbish. Some waste paper is used already to make new paper but most is just thrown away. Far more could be used, especially if people accepted lower-quality paper in packaging and household tissue. Would you do this if you knew you were saving trees?

Rubbers

Natural rubber comes from the sap of the rubber tree which grows particularly in Malaysia, Indonesia and Thailand. There is not enough of it to satisfy the demand for rubber, especially for tyres, so man-made rubbers are essential.

Rubbers are similar chemicals to plastics. They are polymers which can be made from the small molecules which are obtained by cracking naphtha. Propene is one alkene which can be used for this purpose.

Most rubber is used for making tyres, but there are many different sorts of tyre rubber. We are most familiar with tyres on ordinary cars, but many other sorts of tyres are needed. Think of the tyres on huge earth-moving equipment (*figure 4.16*), or the strength needed in the tyres of a 747 jet as it touches down.

Sulphur is an important chemical for making strong tyres. Rubber which has been treated with sulphur to make it strong is known as vulcanized rubber.

Paints and Glues

Paints and glues have a lot in common, as you will know if you have let paint dry on your hands. Using oil, it has been possible to make a whole new range of paints and glues. These substances are polymers, just like the plastics and rubbers. This explains why they dry or set into a solid strong substance.

Polyurethane is one example of a polymer which can be used as a paint (*figure 4.17*). It gives a glossy and water-resistant finish.

Glues which can be made from oil include epoxy resins. Look out for these when you are next in a stationer's shop.

Other Chemicals from Oil

The polymers, which include plastics, rubbers, paints and glues, are not the only chemicals which can be made from oil. Three other groups of chemicals are described elsewhere in this book. One is the group of agricultural chemicals, including fertilizers and pesticides (p.155). Solvents form a second group. *Solvents* are chemicals which dissolve other chemicals (p.133). The third group contains the detergents or man-made soaps (p.199).

Figure 4.16 Huge tyres on an earth-moving vehicle.

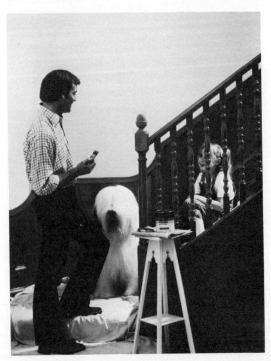

Figure 4.17 Applying a gloss paint, based on chemicals obtained from crude oil.

Chemicals from Oil 73

So many chemicals are made from oil that it is difficult to see how they are all connected to each other. It may help to use the "family tree" in *figure 4.18*. This shows at a glance some of the many substances which are made from crude oil.

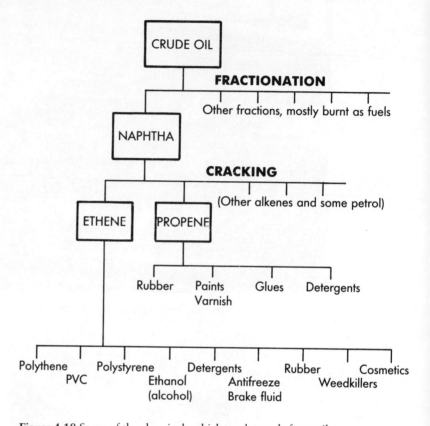

Figure 4.18 Some of the chemicals which can be made from oil.

Questions

1. What are organic chemicals?
2. Draw diagrams, like those in *figure 4.5*, to show how the atoms are joined in the following molecules:
   (a) Chlorine Cl_2
   (b) Hydrogen chloride HCl
   (c) Hydrogen sulphide H_2S
   (d) Ethene C_2H_4
   (e) Carbon dioxide CO_2
3. Draw the structural formulas of all the isomers of pentane C_5H_{12}.
4. Write equations, like those in *figure 4.8*, to show how pentane C_5H_{12} can be cracked to make an alkene.
5. (a) What is a polymer? Explain your answer with a suitable example.
   (b) The chemical tetrafluoroethene can be made into a polymer called Teflon, which is used in non-stick saucepans. The structural formula of tetrafluoroethene is:

   $$\begin{array}{c} F \\ \\ F \end{array} C=C \begin{array}{c} F \\ \\ F \end{array}$$

   Draw the structure of Teflon, showing its repeating unit.
6. Many plastics give off dangerous fumes when they burn. Fumes from these plastics have caused deaths during fires in places like department stores, houses and aircraft. What poisonous gas or gases could be given off by burning PVC, poly(chloroethene)?
7. (a) List 10 objects in your home which have been made from plastics.
   (b) For each object, give one material which could be used to make it instead of plastic.
   (c) From your answers to (b), which materials have plastics replaced and why?
   (d) What problems would arise if plastics could no longer be made?
8. In how many different ways does your life depend on crude oil? Give as many chemical examples as you can.

METALS — **NON-METALS**

I	II											III	IV	V	VI	VII	O
H Hydrogen																	He Helium
Li Lithium	Be Beryllium											B Boron	C Carbon	N Nitrogen	O Oxygen	F Fluorine	Ne Neon
Na Sodium	Mg Magnesium											Al Aluminium	Si Silicon	P Phosphorus	S Sulphur	Cl Chlorine	Ar Argon
K Potassium	Ca Calcium	Sc Scandium	Ti Titanium	V Vanadium	Cr Chromium	Mn Manganese	Fe Iron	Co Cobalt	Ni Nickel	Cu Copper	Zn Zinc	Ga Gallium	Ge Germanium	As Arsenic	Se Selenium	Br Bromine	Kr Krypton
Rb Rubidium	Sr Strontium	Y Yttrium	Zr Zirconium	Nb Niobium	Mo Molybdenum	Tc Technetium	Ru Ruthenium	Rh Rhodium	Pd Palladium	Ag Silver	Cd Cadmium	In Indium	Sn Tin	Sb Antimony	Te Tellurium	I Iodine	Xe Xenon
Cs Caesium	Ba Barium	La Lanthanum *	Hf Hafnium	Ta Tantolum	W Tungsten	Re Rhenium	Os Osmium	Ir Iridium	Pt Platinum	Au Gold	Hg Mercury	Tl Thallium	Pb Lead	Bi Bismuth	Po Polonium	At Astatine	Rn Radon
Fr Francium	Ra Radium	Ac Actinium †															

***58-71 Lanthanum series**	Ce Cerium	Pr Praseodymium	Nd Neodymium	Pm Promethium	Sm Samarium	Eu Europium	Gd Gadolinium	Tb Terbium	Dy Dysprosium	Ho Holmium	Er Erbium	Tm Thulium	Yb Ytterbium	Lu Lutetium
†90-03 Actinium series	Th Thorium	Pa Protactinium	U Uranium	Np Neptunium	Pu Plutonium	Am Americium	Cm Curium	Bk Berkelium	Cf Californium	Es Einsteinium	Fm Fermium	Md Mendelevium	No Nobelium	Lr Lawrencium

Figure 5.1 Metals and non-metals in the Periodic Table.

5 Mankind and Metals

Try to imagine a world without metals. It would be very difficult to make machines and other tools, cars and aeroplanes, suspension bridges and electricity cables without these chemicals.

There are over 60 different metals which occur in the earth. They are all elements, and they make up most of the Periodic Table (*figure 5.1*). Over thousands of years we have discovered how to extract and purify these metals, and we have found uses for them all.

Using Metals

When you think about a metal you probably imagine something strong, hard, shiny and cold to touch. Most metals are like this, although a few are not. For example, the metal sodium can be cut by a knife, and the metal mercury is a liquid at room temperature. Strength and hardness are two of the most useful properties of metals. Machines, tools, bridges and buildings can be made of metals which are strong and hard, because they will last for a long time.

Metals can be made into many different shapes—wires, pipes, blocks and sheets. With a little skill, we can form metals into almost any shape we want.

Metals are good conductors of heat, and are almost the only chemicals which conduct electricity when they are solid. Without metals such as copper, it would be difficult to make wires and other electrical fittings.

Property	Metals	Non-metals
Melting point	Usually high (>200°C).	Usually low (<100°C).
State at room temperature	Solid (except mercury).	Solids with low melting points, liquids or gases.
Appearance	Shiny.	Dull when solid.
Feel	Cold to touch.	Not cold to touch.
Conduction of electricity	Conductors.	Usually non-conductors.
Conduction of heat	Conductors.	Usually non-conductors.
Ease of shaping	Easily worked.	Tend to shatter.

Table 5.1 Differences between metals and non-metals

Mixing metals

Metals can easily be mixed together. A mixture of metals is called an **alloy**. Alloys often have very different properties from the individual metals in them, which is why they can be so useful. Some explanations for this are given on p.144.

Although there are only about 60 metals, there are thousands of possible alloys. They all have slightly different properties which can be exploited by designers, engineers and craftsmen.

Discovering Metals
The earliest humans

The earliest people on this earth used the materials which they found around them. They used stones for making tools, and bones for jewellery. They did not use metals at first, because most metals cannot be found naturally. Only the rare metals gold and silver can be found on their own in the ground. This is because they are unreactive—they do not easily join with other elements to form compounds. Gold and silver are too soft to make into good tools. When they were eventually discovered, there was not much use for them, except for making jewellery (*figure 5.2*) and coins.

Figure 5.2 "Ram in thicket". A gold object from the Ur treasure, nearly 5000 years old (now in the British Museum).

The Bronze Age

All the metals except gold and silver are found in the earth as **ores**. An ore is a chemical compound of a metal, usually mixed with rocks and earth. About 10 000 years ago, people discovered how to get copper from its ore. They did this by heating the ore with charcoal, which is a form of the element carbon. The copper was later mixed with tin producing the alloy known as bronze. Bronze is a hard and strong material, so bronze tools gradually took over from stone.

Copper was one of the earliest metals to be used because it is not very reactive. It takes only a little heating with charcoal to get it from its ore.

The Iron Age

Iron is a more reactive metal than copper. It is therefore more difficult to extract it from its ore. A special furnace has to be made so that the temperature is high enough.

At this high temperature, the iron ore can be made to react with carbon to produce iron. People did not discover how to do this until about 4000 years ago, long after copper had first been extracted.

The hardness of iron made it very useful, and it slowly replaced copper for many purposes. Iron, in the form of steel, is our most widely used metal today.

The modern metals

Metals like aluminium, magnesium and sodium were unknown two hundred years ago. It is only in recent times that we have been able to get them from their ores, because they are so reactive. It is not possible to make them by heating their ores with carbon. Electricity has to be used instead, and electricity was not discovered until the last century.

The more reactive a metal is, the later it was discovered and used by us. Chemists find it useful to put the metals in the order of their reactivity. This order is called the **activity series** (*table 5.2*).

Method of extraction	Metal		Reactivity	Date of discovery
↑ From ore using electricity ↓	Potassium	K	↑	1807
	Sodium	Na		1807
	Magnesium	Mg		1808
	Aluminium	Al	Becoming more reactive	1825
↑ From ore by heating with carbon ↓	Zinc	Zn		Around 0 A.D.
	Iron	Fe		} Iron Age
	Lead	Pb		
	Copper	Cu		} Bronze Age
↑ Found on their own (native) ↓	Silver	Ag		} Earliest civilization
	Gold	Au		

Table 5.2 The activity series

The Metals we Use Today

The earth's crust or **lithosphere** is a mixture of many different substances, including the ores of the metals. Some elements are found quite commonly in these compounds, but most are rare. Just two elements, oxygen and silicon, make up about 75% of the weight of the earth's crust. The metals make up most of the other quarter. Just eight of the elements make up about 99% of the earth's crust (*figure 5.3*).

Two of our most useful metals are also the most common in the earth. These are iron and aluminium. The iron and steel industry is so important to our world today that the whole of the next chapter is about it. The following chapter is about aluminium and the other metals which we use most. These include copper, zinc and lead. Millions of tonnes of these metals are used each year (*figure 5.4*).

How long will our metals last?

The earth's supply of metal ores will not last for ever. We have already used up more metal since 1950 than in the whole history of the world before then. Just like the fossil fuels, the amount of metal ores in our planet is finite. If we use them all up, they cannot be replaced. You often hear talk of an energy crisis. This was explained in part A of the book. We could face a metals crisis as well, which might be more serious. Even if our fossil fuels run out soon, we may be able to fill the energy gap. Perhaps we will use nuclear power (Chapter 2) or perhaps we will find other ways (Chapter 3).

Once we have used up a metal ore, what will replace it? New copper cannot be made from anything except a copper ore.

Metals in short supply

It is always difficult to estimate how long a metal ore will last. New ores are often discovered. New ways of mining may be used. We may also start to use some metals less as they become more expensive.

It seems that we are not in danger at the moment of running out of the most common metals—aluminium, iron and magnesium. With many other metals, the position is different. One estimate of how long some metals may last is shown in *figure 5.5*.

This assumes that our use of the metals increases in roughly the same way as it has over the past 30 years. These figures look alarming and many people do not agree with them. For example, there have been two big discoveries of zinc and lead ores recently. These should keep us supplied into the next century.

Even if the figures are not completely right, we should still be worried. We should take care to make our metals last as long as possible. It would not be easy to replace metals like copper, tin, zinc and lead.

Recycling

Even if a metal is quite common in the earth's crust, it does not mean that it is easy to obtain the metal. Some ores are quite rich in the metal, others are poor. Large amounts of material may have to be mined just for a small amount of metal. This is obviously

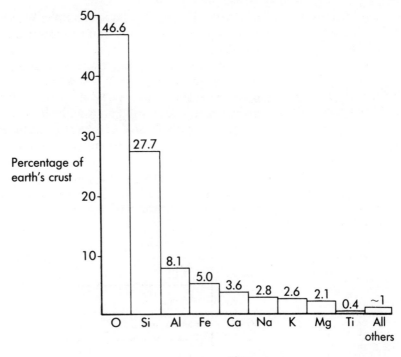

Figure 5.3 The abundance of elements in the earth's crust.

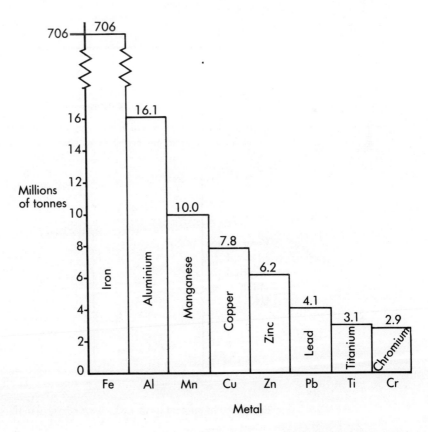

Figure 5.4 The world's production of the most commonly used metals (1980). Notice how much iron is made compared with all the other metals.

Mankind and Metals 81

expensive. Copper is a good example. We seem to have run out of rich ores, and we are now mining ores which contain only 0.5% of metal.

As an ore becomes scarce and more difficult to mine, the price of the metal tends to go up. We should expect most metals to become more expensive as time goes on. This will hurt the poorer countries most, because they will find it difficult to pay the high prices. If the metal is more expensive, it is worth mining poorer ores. When the price of tin increased in recent years, some old Cornish tin mines were reopened.

If a metal is expensive, people think twice about using it or throwing it away. It often becomes cheaper to collect old bits of metal and recycle them than to use new ores. Scrap metal is already

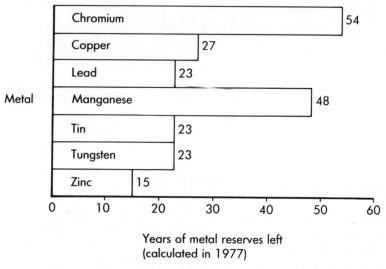

Figure 5.5 Some metals which may run out soon. Even if most of them last much longer, which they probably will, we are living dangerously.

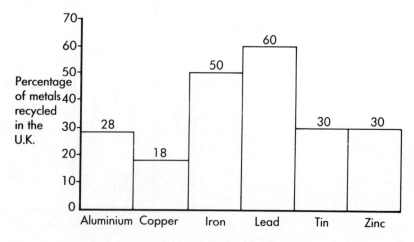

Figure 5.6 Important metals which are recycled in the U.K.

being recycled in large quantities by the scrap industry (*figure 5.6*).

The savings can be enormous. Producing copper from scrap costs only 3% as much as making it from its ore. Recycling aluminium in the U.K. saves as much energy as the whole farming industry uses in England and Wales each year.

Scrap metal from industry is quite easy to recycle, because it is in large amounts. It is more difficult to recycle metals which we use and throw away as individuals. All the pieces have to be collected together, and this is expensive.

Recycling is most easy if we throw away things made of pure metals. For example, 50% of copper is recycled, as is 80% of lead from batteries. Objects like "tin" cans (steel coated with tin) are more difficult, but it is now possible to recycle these too.

Figure 5.7 Recycling aluminium. It is expensive to make new aluminium from its ore, because so much energy is needed (p.101). Only 5% of this energy is required to make the same amount of aluminium from scrap. Recycling aluminium will save energy as well as aluminium.

6 Iron and Steel

Our modern world is literally built on steel. Most of our large buildings are constructed round a steel frame or contain steel-reinforced concrete (*figure 6.1*).

Transport in most parts of the world is based on steel. Steel makes up nearly 75% of the weight of an average car. We also use steel for making lorries, ships, trains and railway lines. Machines based on steel are used in all our important industries. Steel tools are used to shape glass, crush stone, mix concrete, make plastics, and form metals. We use 50 times more steel in the world than any other metal.

Steel is not a pure metal. It is an alloy. It usually contains at least 95% of iron, but different steels contain different amounts of many other metals. Since steel is always made from iron, this chapter starts with the metal iron itself.

Making Iron

Iron is the second most common metal in the earth's crust. It is quite a reactive metal, so it is found combined with other elements in ores.

Iron ores

The two common ores of iron both contain iron oxides, compounds of iron and oxygen. These are haematite Fe_2O_3 and magnetite Fe_3O_4. Magnetite gets its name because it is magnetic, like iron itself.

There seems little danger that we will run out of these ores for a long time. Even so, it makes sense to use them wisely. The U.S.S.R. and Australia are the two biggest producers of iron ore. The U.K. used to be a major producer of iron ore, but most of the iron ore now used in the U.K. is imported.

Extracting the iron

Iron ore is a mixture of many different chemical compounds. Iron oxide makes up about 60% of most ores, although some ores which are being used today contain as little as 40%. Most of the rest of the ore is made up of compounds of silicon and oxygen.

Figure 6.1 The Kennedy Space Centre, U.S.A. The main building, constructed round steel, is the world's largest man-made structure.

This is to be expected, since silicon and oxygen are the two most common elements in the earth's crust.

In order to obtain the iron, two things have to be done. The impurities including silicon dioxide have to be removed, and the iron oxide must be chemically changed into iron. The impurities are removed by heating the ore with limestone (calcium carbonate). The iron is obtained by heating the ore with carbon in the form of coke. These reactions take place together in a huge tower called a blast furnace. This method of making iron is called *smelting*.

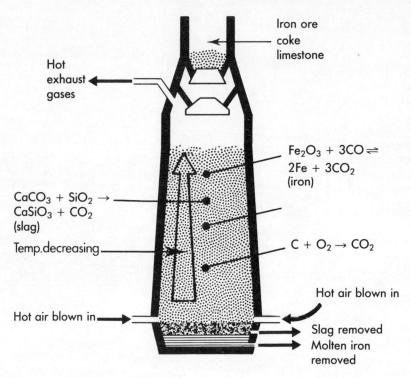

Figure 6.2a The blast furnace.

Chemistry inside the blast furnace

While reading this section, you should keep referring to the diagram in *figure 6.2a*.

A modern blast furnace is a steel structure over 70 m high. The reactions take place inside a steel cylinder which is 30 m high. Above this cylinder is the machinery used to fill up the furnace, and pipes to take away the gases which are produced. The blast furnace is used twenty-four hours each day for several years, until the lining of the furnace wears out. Modern furnaces can produce 8000 tonnes of iron each day.

The blast furnace is loaded from the top with iron ore, coke and limestone. A special loading system is used so that little pollution can escape from the furnace. About one tonne of coke is needed for each tonne of iron which is made.

Hot air is blown in at the bottom of the furnace. The coke, which is mostly carbon, reacts with the oxygen in the air to form carbon dioxide:

Carbon + Oxygen → Carbon dioxide
$C + O_2 \quad\quad CO_2$

This reaction is exothermic. It helps to heat the furnace.

The carbon dioxide gas starts to rise up the furnace, away from the hot air. As it passes more hot coke, it reacts with it to form carbon monoxide:

86 Chemistry in Use

Figure 6.2b A blast furnace at the Port Talbot works, South Wales. The loading system, taking raw materials from the pile to the top of the furnace, can be seen clearly in the photograph. The furnace itself is just below the huge pipes in the middle of the picture. The hot exhaust gases are carried away through these pipes. Heat from the gases is used in other parts of the works. The gases are cleaned before being released into the atmosphere.

$$\text{Carbon} + \text{Carbon dioxide} \rightarrow \text{Carbon monoxide}$$
$$C + CO_2 \quad\quad\quad 2CO$$

Carbon monoxide is the chemical which reacts with the iron oxide to form iron:

$$\text{Iron(III) oxide} + \text{Carbon monoxide} \rightleftharpoons \text{Iron} + \text{Carbon dioxide}$$
$$Fe_2O_3 + 3CO \quad\quad\quad 2Fe + 3CO_2$$

In this reaction, oxygen is being taken away from the iron oxide. The iron oxide is being reduced. A **reduction** is the opposite of an oxidation (p.8). The carbon monoxide which takes part in this reaction is called a **reducing agent**. A reduction cannot take place without an oxidation, because if one chemical loses oxygen then another one must gain it. Reactions like this are called **redox reactions**. In the blast furnace, the carbon monoxide is oxidized while the iron oxide is reduced.

The iron which is formed in this reaction is so hot that it is a liquid. Liquid iron runs down the furnace and collects at the bottom. It can be tapped off and solidified into blocks called "pigs", which give it the name pig-iron.

While these reactions are happening, the limestone reacts with the impurities to form a liquid slag. A layer of slag settles on top of the pig-iron. It is tapped off from time to time:

Calcium carbonate + Silicon dioxide → Calcium silicate + Carbon dioxide
(limestone)   (impurity)   (slag)
$CaCO_3 + SiO_2 \rightarrow CaSiO_3 + CO_2$

The whole process runs continuously, hour after hour and day after day. Iron ore, coke and limestone are fed into the top, while iron and slag are removed from the bottom.

It is important to save energy, which saves money, whenever possible. For this reason, the hot gases escaping from the top are used to heat up the air going in at the bottom. They are also used to heat the factory buildings and offices.

Pig-iron

The pig-iron which comes out of the furnace contains about 93% iron. The other 7% are impurities, including 4% of carbon. This makes the iron very weak and brittle, so that it shatters easily.

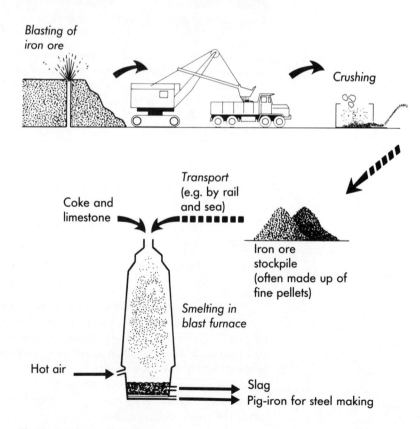

Figure 6.3 Extraction of iron.

Pig-iron cannot be used for making many objects directly, except for things like grates and railings which do not need to be very strong.

Most pig-iron is immediately turned into a much more useful metal—steel.

Making Steel

Industrialized countries use large amounts of steel. Most steel is made in the developed Western world, and in the U.S.S.R. Japan and the U.S.A. are big producers (*figure 6.4*).

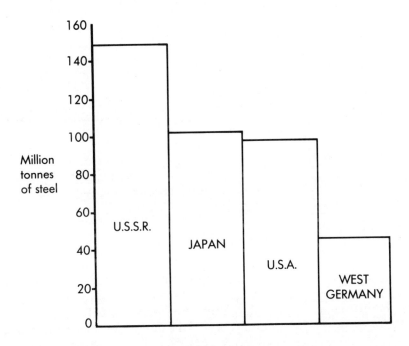

Figure 6.4 The main steel producers in 1980. These four countries produced over half the world's steel between them.

Steel is a good measure of prosperity or economic trouble. Steel production in the U.K. fell by half when industries went through difficulties in the early 1980s.

Many Western countries, including the U.K., have been making steel for about 100 years. They are starting to run out of their own ores. New steel developments in the U.K. are on the coast, so that ores can be shipped to them. Many developing countries are now building their own steel works, near to their own ores.

Turning pig-iron into steel

The main impurities in pig-iron are carbon, silicon and phosphorus. Most of the carbon and all the silicon and phosphorus have to be removed, otherwise the iron is weak and brittle. The carbon is often removed completely at the start along with the

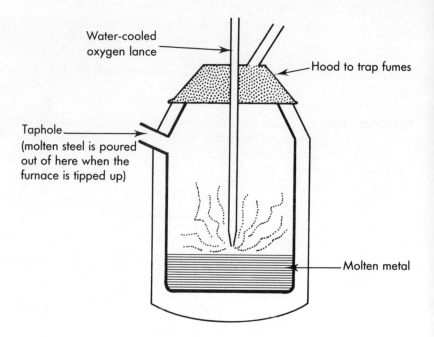

Figure 6.5a The basic oxygen furnace. Pure oxygen is blown at the molten iron to oxidize the impurities. The oxidized impurities are blown out or removed as a slag.

silicon and phosphorus. The right amount of carbon can then be added back to make the steel which is wanted.

In many steel factories, the pig-iron is not even allowed to cool and solidify. This would waste heat. Instead, it goes directly to a furnace where the impurities can be burnt off by reacting them with oxygen.

Most steel is made today by the basic oxygen process (*figure 6.5a, b*). Pure oxygen is blown into the molten pig-iron. The impurities are oxidized, forming chemicals like carbon monoxide, silicon dioxide and phosphorus oxide. These chemicals can be removed quite easily. Carbon monoxide gas simply leaves the furnace. The other oxides can be removed as a slag. Steel-making is another example of an oxidation.

A modern steel furnace can make 300 tonnes of steel in only 35 minutes. Up to 25% of scrap metal can be used in the basic oxygen furnace. This helps to recycle the metal and cuts the cost of steel-making.

Making different steels

The liquid steel is usually allowed to cool down slowly. This gives a strong steel which is also malleable (easily worked). If the steel is cooled down quickly, a hard steel is made. This treatment is called *quenching*.

All steels contain a small amount of carbon. Mild steel, which

Figure 6.5b A 300 tonne Basic Oxygen Steelmaking furnace at the British Steel Corporation works in Scunthorpe. Molten iron is being poured into the furnace from a giant container.

contains about 1% of carbon, is quite soft and ductile (easily drawn into shape). If more carbon is added, the steel becomes much harder (*table 6.1*). Steel also becomes more brittle as carbon is added, which means that it shatters more easily. It can be made tougher by heating it and then allowing it to cool again. This treatment is called *tempering*.

Type of steel	Percentage of carbon	Uses
Soft	Up to 0.15%	Sheets and wires
Mild	0.15 to 0.25%	Building and general engineering
Medium-carbon	0.2 to 0.5%	Strong springs
High-carbon	0.5 to 1.4%	Hammers and chisels

Table 6.1 Steels containing different amounts of carbon

Other types of steel can be made by mixing the iron with small amounts of other metals. Many of these metals are quite rare and are only found in one or two parts of the world. They belong to a group of metals called "strategic" metals. The word "strategic" is used because these metals make alloys which are vital to the economies and military forces of most countries.

Iron and Steel 91

Figure 6.6 Iron and steel making is big business. The photograph is an aerial view of the Port Talbot works in South Wales.

The strategic metals

Cobalt, molybdenum, niobium, tantalum and vanadium are not exactly household words. Even so, our lives would be seriously affected if we could not get hold of these metals.

The metals manganese, niobium, molybdenum and chromium are used to harden steel. If supplies of these dried up, we would not be able to make things like cars or industrial machinery.

These and other metals are also needed for military reasons. Tungsten steels are used for armour plating, while cobalt is used in jet engine blades. Titanium is an important metal for military aircraft and other vehicles (*figure 6.7*).

Information about some of these strategic metals is shown in *table 6.2*.

Now look at which countries produce these metals. Western countries depend on other countries, including S. Africa and the U.S.S.R., with which they are not very good friends. You can see the point of the advertisement from the S. African Embassy shown in *figure 6.8*.

92 Chemistry in Use

Imagine what would happen if the U.S.S.R. or S. Africa stopped supplying these metals. For example, without platinum catalysts, much of our chemical industry would grind to a halt.

Many countries are starting to buy these metals and store them away, just in case supplies are cut. The U.S.A. now has about three years worth of supplies and the U.K. is also building up stocks.

Metal	Main producer (% of world total)	Uses
Chromium	S. Africa 35, U.S.S.R. 25	Stainless steel
Cobalt	Zaire 50, Zambia 10	Jet engines
Manganese	U.S.S.R. 38, S. Africa 20	Hard steels
Molybdenum	U.S.A. 62, Chile 13	Steels
Niobium	Brazil 79	Steel cutting tools
Platinum	U.S.S.R. 48, S. Africa 46	Catalysts
Tantalum	Thailand 44, Malaysia 14	Electronics
Titanium	U.S.A., Australia	Aircraft
Tungsten	China 26, U.S.S.R. 18	Steel armour plating
Vanadium	S. Africa 36, U.S.S.R. 29	Steels

Table 6.2 Some strategic metals

Figure 6.7 The Blackbird, the fastest aircraft in the world. Titanium makes up 90% of this aircraft. Titanium is lighter than steel, although not as light as aluminium. It makes strong alloys and resists corrosion well.

Iron and Steel 93

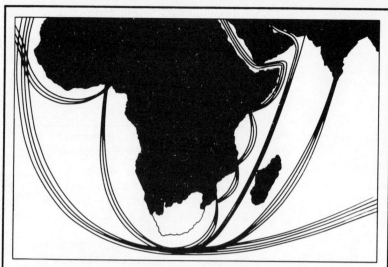

Figure 6.8 The importance of South Africa. Despite South Africa's importance, most countries will not sell arms to South Africa, as the advertisement says. This is because they disapprove of South Africa's internal policies.

Rusting

There is one big problem with iron and steel—they rust (*figure 6.9*). Rusting destroys 20% of the world production of these metals each year. The cost of this is well over £10 billion.

Steel is such a useful and relatively cheap material that we have learnt how to cope with rusting. We have invented many different ways of slowing it down and making our steel objects last longer.

The chemistry of rusting

Most metals react with chemicals in the air, including oxygen, water and the acids caused by pollution. The metals are slowly

Figure 6.9 A scrap heap. The scrap industry recycles metals. Recyling saves metals. It also saves money, because countries do not need to buy so much new metal from abroad.

eaten away by these reactions. Chemists call this **corrosion**. The rusting of iron and steel is an example of corrosion.

Rusting only happens if oxygen and water can get at the iron or steel together. Ordinary air always contains both oxygen and water vapour, so iron and steel rust in the open air. Oxygen and water react with the iron to form a hydrated iron oxide, which we call rust. "Hydrated" means simply that there is water present. A simple equation for this can be written:

Iron + Oxygen + Water → Hydrated iron oxide (rust)

The layer of rust is very weak. It soon comes off, and fresh iron underneath starts to rust. Eventually a whole piece of metal can corrode right through.

Rusting will not happen if air and water are kept away from the surface of the metal. This means that rusting can be prevented by covering the surface of the metal, to keep out air and water.

Rust prevention using covering layers

One of the easiest ways to stop rust is to paint the surface of the metal. Many large steel objects, including bridges, are protected in this way. The paint wears out after a while and repainting is necessary.

Everyone is familiar with the rusting of cars, lorries and other vehicles. The underneath of a car is especially likely to rust. This is

Iron and Steel 95

Figure 6.10 Layers of paint stop car bodies from rusting too quickly. Automatic machines are used to give a complete and even covering layer.

because it gets chipped by stones and splashed by water from the road. If the roads have been salted, rusting is even faster. People often cover parts of the underneath with grease or with a type of plastic called underseal. This slows down the rusting but does not stop it for ever.

One other simple method of covering is to use a layer of tin. The steel is dipped into a bath of liquid tin, which forms a layer over the steel. This is called *tin-plating*. Many of the food cans which we call tin cans are made from steel which is tin-plated. Tin-plating works well as long as the can is not damaged. As soon as the layer of tin is chipped or cracked, the steel underneath will start to rust again.

Rust prevention using the activity series

Iron is a reactive metal, but metals like zinc and magnesium are even higher in the activity series. These metals can be used to protect the iron from rusting. They will corrode instead of the iron, because they are more reactive. The iron itself remains untouched.

A common way to do this is to coat steel with zinc. This can be done by dipping the steel in liquid zinc, which is called *galvanizing*. Steel coated with zinc is called galvanized steel. Even if the layer of zinc is scratched or broken, the steel underneath is still protected. The zinc corrodes away first, leaving the steel free from rust. Galvanizing is such a good way of stopping rust that huge amounts of steel are protected by it. About a third of all the zinc made in the world is used for galvanizing steel.

A second way of stopping rust by using the activity series is called *sacrificial protection*. Lumps of reactive metals like zinc or magnesium are sacrificed to save the steel. Big steel objects like ships' hulls or piers are protected by attaching lumps of these metals to them. From time to time, new pieces of zinc or magnesium must be put on to replace the corroded ones. Thousands of miles of gas and water pipes are protected in the same way. It is much cheaper to dig up a few lumps of magnesium than to dig up miles of pipes.

Rust prevention using alloys

Most steels will rust but it is possible to make special steels, called stainless steels, which do not rust. *Stainless steel* is an alloy of iron with about 10% of chromium and some nickel. It is more expensive than ordinary steel, which is one reason why it is not often used for making cars.

A familiar use of stainless steel is for making cutlery, which would be unpleasant if it rusted.

Questions

1 (a) Give two important chemical differences between pig-iron and mild steel.
   (b) Explain two methods which can be used to change the properties of steel.
2 (a) Explain the chemical reaction which takes place when iron rusts.
   (b) Describe three methods which can be used to prevent the corrosion of iron.
3 (a) What advantage is there in making cars out of galvanized steel?
   (b) Suggest reasons why cars are not usually made of this material.
4 Some information about the corrosion of galvanized steel is given in the following table:

Site	Type of atmosphere	Speed of corrosion (microns per year)
Llanwrtyd Wells	Rural	2
Calshot, Hants	Marine	3
Motherwell, Scotland	Industrial	5
Woolwich, Kent	Industrial	4
Sheffield (University)	Industrial	5
Sheffield (Attercliffe)	Industrial	15
Basrah, Iraq	Dry, sub-tropical	0.3
Apapa, Nigeria	Marine tropical	0.8
Congella, Durban	Marine industrial	5

(a) Suggest reasons for the differences between tropical, rural, marine and industrial areas.
(b) Suggest a possible reason for the big difference between the two areas in Sheffield.

5 (a) List the raw materials used for making iron.
(b) Explain, step by step, the chemical reactions which occur inside a blast furnace.

6 Explain, with examples, why steel-making involves oxidation reactions.

7 (a) Why are some metals called "strategic metals"?
(b) Chromium, manganese and vanadium are strategic metals. Which countries are the main producers of these metals?

7 Aluminium and Other Metals

Using and Making Aluminium (Al)

Using aluminium

Aluminium was a rare and expensive metal only 100 years ago. It used to cost more than gold. Today we use more aluminium than any other metal except iron.

Aluminium is a light metal—the same volume of steel weighs three times as much. It does not corrode easily and it is quite strong, especially when it is made into alloys. These properties explain why aluminium is useful for making energy-saving lightweight vehicles, including aircraft, buses and ships. For the same reasons it is used in modern buildings, from window frames to whole faces of office blocks. At home you can find washing machines or freezers made out of this modern metal, together with cooking foil, milk bottle tops and other sorts of packaging (*figure 7.1*).

Figure 7.1 Aluminium in food packaging. Can you spot the use of aluminium in each of these packages?

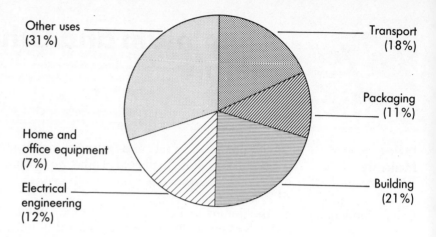

Figure 7.2 The uses of aluminium in the U.K. (1980).

The main disadvantage of aluminium is its price. It costs about six times as much to produce aluminium than to produce steel.

Aluminium is one of the best conductors of electricity, together with the more expensive metals silver and copper. It is ideal for making thick overhead power cables because it is also light and does not put too much strain on the pylons. The cables themselves need to have strong steel cores, because they would otherwise break under their own weight.

The main areas in which aluminium is used are shown in *figure 7.2*.

Alloys of aluminium

Alloys of aluminium usually contain at least 95% aluminium. Small amounts of other metals can increase strength and corrosion resistance, and make the metal more malleable.

In general, copper and zinc are used to make high-strength alloys. Magnesium and silicon give pleasant-looking, corrosion-resistant alloys, used for making windows and doors.

The high-strength alloys used for aircraft contain copper, with small amounts of zinc, magnesium and silicon. The name duralumin is sometimes used for this type of alloy.

Bauxite—the aluminium ore

Bauxite is the name given to the common ore of aluminium. It is mostly aluminium oxide Al_2O_3, although it contains impurities like iron oxide.

Bauxite is found in various countries, including Jamaica in the Caribbean, and West African countries like Guinea and Ghana. It might seem an advantage for developing countries to have rich supplies of an ore like bauxite, because they could sell it to industrialized countries. This is not always the case.

In Ghana, a huge hydroelectric scheme was built on the Volta River together with an aluminium factory, owned by an American company. The Ghanaians wanted local bauxite to be used, but the American company refused to use it. This means that the Ghanaians do not get the benefit of their own ore. They get electricity from the Volta River dam, but not very cheaply, because the American company pays much less for its electricity that it would in a Western country. In addition, to make the Volta dam, 5% of the country had to be flooded and 80 000 people were resettled. Imagine the effect of this in your area.

The price of bauxite fell in the early 1980s, because of the world recession. This badly affected many countries, including Jamaica. 75% of Jamaica's exports are connected with bauxite. The falling price of bauxite led to strikes and unemployment of at least 25% of the work force.

Extracting aluminium

Aluminium is a reactive metal. It cannot be made by heating its ore with carbon, in the way that iron is made. Electricity has to be used instead. Pure aluminium oxide is needed for this, so the bauxite has to be purified first. The pure aluminium oxide is known as alumina.

It takes enormous amounts of electricity to make aluminium from alumina. This is why aluminium is quite an expensive metal.

Many of the factories, which are called smelters, are built near supplies of hydroelectric power. These include the smelter in Ghana, and others in the mountains of Canada and Scotland (*figure 7.3*). About half the world's aluminium is made using hydroelectric power. Other smelters must be near supplies of fossil

Figure 7.3 The aluminium smelter owned by British Alcan Aluminium at Fort William, Scotland. The pipes on the side of the mountain are part of the hydroelectric scheme which makes electricity for the factory.

Aluminium and Other Metals 101

fuels. The smelter at Lynemouth in Northumbria was built next to a big coalfield, and even has its own power station.

The method used to extract aluminium from aluminium oxide is called **electrolysis**. It takes place in large tanks called **electrolysis cells**. The electricity is passed into the alumina and out again through **electrodes**. One electrode is connected to the positive terminal of the electricity supply. This is the positive electrode,

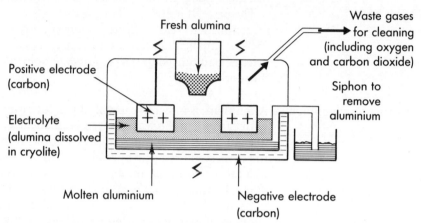

Figure 7.4 Electrolysis of alumina (aluminium oxide) dissolved in cryolite.

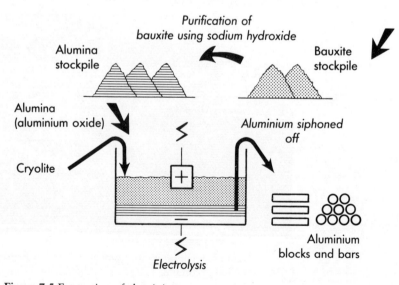

Figure 7.5 Extraction of aluminium.

sometimes called the **anode**. The negative electrode is connected to the negative terminal of the electricity supply. It is known as the **cathode**.

Alumina (aluminium oxide) is a solid which does not normally conduct electricity, because it is not a metal. It can be made to conduct by dissolving it in a liquid called cryolite at about 950°C. Cryolite is also a compound of aluminium, but it is not used up during electrolysis.

When the alumina conducts electricity, it is split up into its elements, aluminium and oxygen (*figure 7.4*). Liquid aluminium collects at the bottom of the electrolysis cell next to the negative electrode. It is siphoned off from time to time. It can be made into alloys as required, and then turned into sheets, tubes and other products. Oxygen is made at the positive electrode, but it is not the only chemical which is produced there. The electrode is made of carbon, so it slowly burns away in the oxygen at the high temperatures which are used. Carbon dioxide is produced, and the electrodes have to be replaced when they have burnt away. The making of aluminium is summarized in *figure 7.5*.

In order to understand more fully how aluminium is made by electrolysis, it is necessary to look more closely at the connection between chemicals and electricity.

Chemicals and Electricity

Chemicals which can conduct electricity are called **conductors**. The only chemicals which are good conductors of electricity when they are solid are the metals and graphite (a form of carbon). Silicon conducts electricity, but less easily. It is known as a **semiconductor**. Silicon is important in the electronics and computer industries and in solar batteries.

Some substances which are not normally conductors can be made to conduct electricity if they are melted or dissolved in a liquid. These chemicals are called **electrolytes**. Aluminium oxide is an example of an electrolyte. Solid aluminium oxide does not conduct electricity, but it will conduct if it is dissolved in liquid cryolite.

Other substances cannot be made to conduct electricity even if they are melted or dissolved in a liquid. They are called **non-electrolytes**. Many non-electrolytes are useful as electrical **insulators**, because they do not allow an electric current to pass. Plastics, rubber and ceramics are non-electrolytes which are often used as insulators around wires and other electrical products.

How electrolytes conduct electricity

It is helpful to imagine that electricity is conducted whenever particles with electric charges move. In metals, it is the small negatively charged electrons which carry the electric current (p.144).

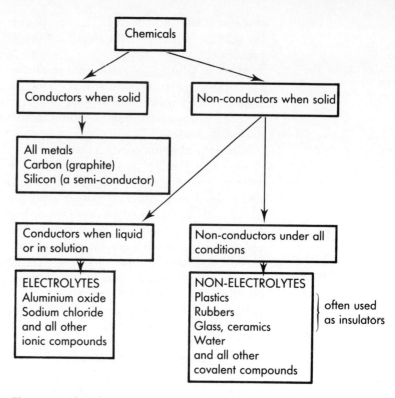

Figure 7.6 Classifying conductors and non-conductors.

Since electrolytes can conduct electricity, it seems that they must contain particles which have electric charges. These particles are called **ions**. Electrolytes are known as *ionic compounds*, because they are compounds made of ions.

In a solid electrolyte, the ions are fixed in place and cannot move. If the ions cannot move, then electricity is not conducted. This is because electricity is only conducted when particles with electric charges, in this case ions, are moving. As soon as the electrolyte is melted or dissolved in a liquid, the ions are free to move. This means that electricity can now be conducted (*figure 7.7*).

More about ions

All atoms normally contain a certain number of positive charges (protons), balanced by an equal number of negative charges (electrons). The positive charges cancel out the negative charges, so an atom has no electric charge overall. Ions are atoms, or groups of atoms, which have gained extra electrons or lost electrons.

Atoms which gain electrons form **negative ions**. Negative ions have more electrons than protons. An example of a negative ion is the oxide ion in aluminium oxide. An oxide ion has two more electrons than an oxygen atom, but it has kept the same number of protons. This means that it contains two extra negative charges, and, to show this, the symbol for the oxide ion is O^{2-} (*figure 7.8*).

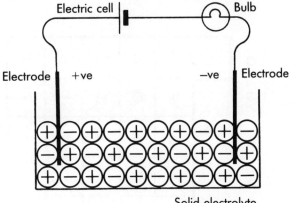

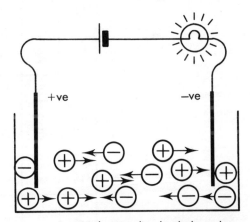

Figure 7.7 Conduction of electricity by electrolytes.

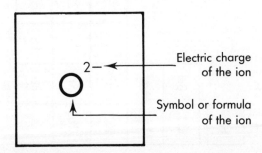

Figure 7.8 Symbol for the oxide ion.

Aluminium and Other Metals 105

I	II												III	IV	V	VI	VII	O
H⁺																		He
Li⁺	Be												B	C	N	**O²⁻**	**F⁻**	Ne
Na⁺	**Mg²⁺**												**Al³⁺**	Si	P	**S²⁻**	**Cl⁻**	Ar
K⁺	**Ca²⁺**	Sc	Ti	V	Cr	Mn	**Fe²⁺/³⁺**	Co	Ni	**Cu²⁺**	**Zn²⁺**	Ga	Ge	As	Se	**Br⁻**	Kr
Rb	Sr	Y	Zr	Nb	Mo	Tc	Ru	Rh	Pd	Ag	Cd	In	Sn	Sb	Te	I⁻	Xe
Cs	Ba	La	Hf	Ta	W	Re	Os	Ir	Pt	Au	Hg	Tl	**Pb²⁺**	Bi	Po	At	Rn

Figure 7.10 Ions and the Periodic Table. What patterns can you see here?

Atoms which lose electrons form **positive ions**. Positive ions have fewer electrons than protons. In other words, they have more positive charges. An example of a positive ion is the aluminium ion in aluminium oxide. An aluminium ion is an aluminium atom which has lost three electrons, but has kept the same number of protons. This means that it has three more protons than electrons, so it has three positive charges overall. The symbol for the aluminium ion is Al^{3+} (*figure 7.9*).

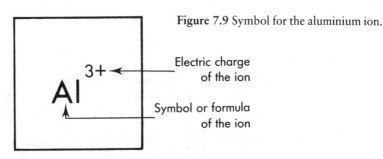

Figure 7.9 Symbol for the aluminium ion.

Ions and the Periodic Table

The ions which are formed by some common elements are shown in part of the Periodic Table in *figure 7.10*.

There is a clear pattern. Metals and hydrogen form positive ions, while non-metals form negative ions. This can be explained by looking at the arrangement of the electrons in these atoms and ions. The arrangement of electrons in atoms has been described on pages 59 to 60. You should read this section now if you have not already done so.

The noble gases, in the right-hand column of the Periodic Table, do not form ions. Indeed, they rarely form compounds at all. This is explained by saying that noble gas atoms have full electron shells, which are stable (p.60). Noble gas atoms do not lose or gain electrons to form ions.

The elements in the first column, including hydrogen and sodium, form ions with a single positive charge. Atoms of these elements have just one electron in their outer electron shell. If they lose this one electron, they then have stable electron arrangements, like noble gases. This means that they have one fewer electron than protons, so they have a single positive charge overall.

The elements in the second column, including magnesium, form ions with two positive charges. Atoms of these elements have two electrons in their outer shells. They lose them both, which gives them two positive charges overall.

In the same way, aluminium forms an ion with three positive charges, because each aluminium atom loses its three outer electrons.

Carbon does not form an ion with four positive charges, as you might expect. The reason is that it takes too much energy to remove four negatively charged electrons from the atom—they are held in by the positively charged nucleus. Carbon forms covalent compounds instead (p.60).

The elements in the seventh column, including chlorine, form ions with a single negative charge. Atoms of these elements have seven electrons in their outer shells. They gain one electron to give them the stable arrangement of the noble gases, so they have an extra negative charge.

Elements in the same column as oxygen form ions with two negative charges. They gain two electrons to reach the noble gas arrangement.

Ions with three or four negative charges are rarely formed,

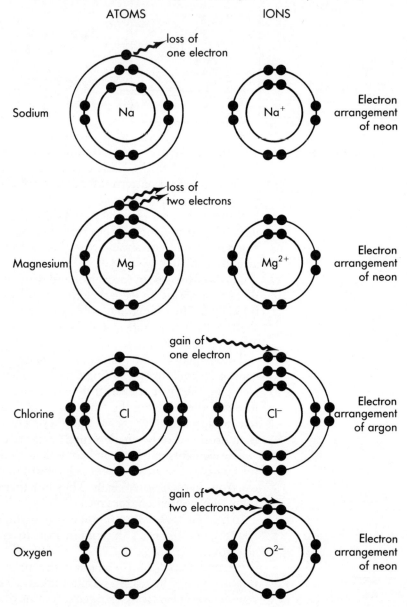

Figure 7.11 Formation of ions from atoms.

Explaining the electrolysis of aluminium oxide

because these negative charges will repel each other. The attraction to the nucleus is not strong enough to hold them all in.

The way in which some atoms form ions is shown in *figure 7.11*.

When electrolytes conduct electricity, they are always chemically changed. Pure aluminium oxide is a white powder. When it conducts electricity, it is chemically changed into aluminium metal and oxygen gas. What has happened is that aluminium ions and oxide ions have been turned back into atoms of aluminium and molecules of oxygen.

Aluminium ions are positively charged. They are attracted to the negative electrode, because opposite electric charges attract each other. Each aluminium ion is given three electrons at the negative electrode. These electrons cancel out the three positive charges of the aluminium ion. This turns the aluminium ion into an ordinary atom of aluminium metal. Aluminium is therefore formed at the negative electrode during electrolysis (*figure 7.12*).

This can be described by an equation:

Aluminium ions + Electrons → Aluminium atoms

The symbol for an electron is e^-, so the equation can be written:

$$Al^{3+} + 3e^- \rightarrow Al$$

Oxide ions are negatively charged, so they are attracted to the positive electrode. Here they lose their extra electrons to become the element oxygen again:

Oxide ions − Electrons → Oxygen molecules

Oxygen molecules contain two atoms, so it takes two oxide ions to form one molecule of oxygen. Each of these oxide ions must lose its two extra electrons:

$$2O^{2-} - 4e^- \rightarrow O_2$$

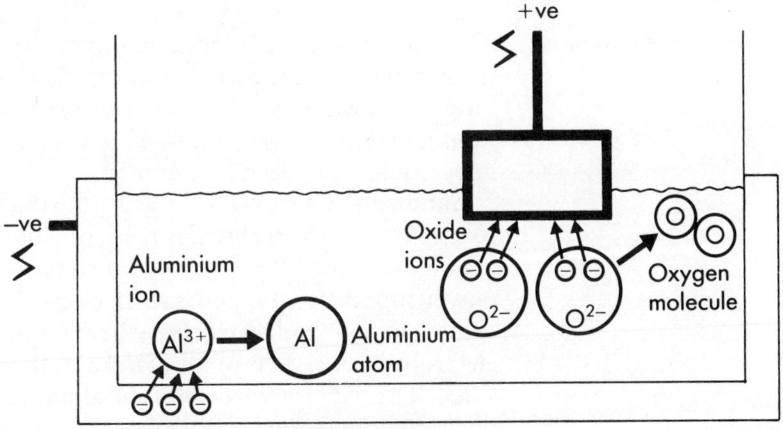

Figure 7.12 Electrolysis of aluminium oxide.
Oxide ions lose electrons to the positive electrode.
Aluminium ions gain electrons from the negative electrode.

Aluminium and Other Metals

Aluminium—a Wolf in Sheep's Clothing

One reason aluminium is so useful is that it does not corrode easily. The biggest problem with iron and steel is that they react with water and oxygen in the air and rust away.

Aluminium is higher in the activity series than iron (p.79), so it might be expected to corrode more easily. Aluminium does indeed react quickly with oxygen in the air, forming a coat of aluminium oxide on the metal. This coat sticks to the metal and protects it, unlike the rust on iron, which just flakes off. Even if the surface of the aluminium is scratched, the exposed metal reacts immediately to form a fresh coat of aluminium oxide.

Using the reactivity of aluminium

The reactivity of aluminium is used in a special method of welding steel called thermit welding. A mixture of aluminium and iron oxide is placed between the two pieces of steel to be welded. The reaction is started by lighting a fuse of magnesium. When the reaction starts, the aluminium takes the oxygen away from the iron oxide, because aluminium is more reactive than iron. Aluminium oxide and iron are produced. The reaction is exothermic enough to melt the iron.

As the iron cools down and solidifies, it welds the two pieces of iron together:

Aluminium + Iron oxide $\rightarrow$ Iron + Aluminium oxide
$$2\,Al + Fe_2O_3 \quad\quad 2Fe + Al_2O_3$$

The reactivity of aluminium can also be used to extract a metal like chromium from its ore. Aluminium is more reactive than chromium, so chromium can be made by heating aluminium with chromium oxide:

Aluminium + Chromium oxide $\rightarrow$
$\quad\quad\quad\quad\quad\quad\quad\quad$ Chromium + Aluminium oxide

Anodized aluminium

The thickness of the layer of aluminium oxide is often increased deliberately, to give extra protection against corrosion. If an aluminium window salesman or shower unit salesman calls at your house, look at the brochures. You will almost certainly see them advertising "anodized" aluminium.

Anodizing is a way of increasing the thickness of the aluminium oxide layer by electrolysis. A piece of aluminium is made into a positive electrode by connecting it to the positive terminal of a power supply. It is dipped into an electrolysis cell containing dilute sulphuric acid, and an electric current is passed through it. During electrolysis, oxygen is made at the aluminium electrode (*figure 7.13*). The oxygen reacts with the aluminium, increasing the layer of aluminium oxide.

As well as giving extra protection, the thicker layer of aluminium oxide also gives a smart finish to aluminium products (*figure 7.14*). The anodized layer can even be dyed if required.

Zinc (Zn)

Zinc is quite a reactive metal. It lies between aluminium and iron in the activity series (p.79), so it is quite difficult to extract zinc from its ores.

Zinc was discovered as a pure metal about 2000 years ago, and it has also been used in alloys for hundreds of years. For example, the Romans made and used the alloy called brass, which contains about 60% copper and 40% zinc.

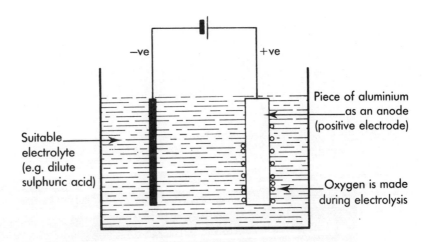

Figure 7.13 Anodizing aluminium.

Figure 7.14 Aluminium saucepans. Anodized aluminium can be left with its shiny finish or it can be dyed. Alternatively, aluminium can be coated with enamel or other substances to give a variety of finishes.

Using zinc

The most important use of zinc is for coating steel to prevent rusting (p.96). Zinc corrodes ten to fifty times more slowly than steel and a thin layer of zinc can protect the steel very well. The layer of zinc is usually put on by galvanizing (dipping the steel in molten zinc). Galvanizing is not always suitable, so other methods are used. These involve electroplating or spraying molten zinc onto the steel. Most zinc-coated steel is used in the building industry (*figure 7.15*).

Figure 7.15 The Tasman bridge, Australia, constructed from galvanized steel. The zinc coating protects the steel underneath from rusting.

About a quarter of our zinc is turned into castings of zinc alloys. A casting is a lump of metal shaped for a particular purpose. Zinc is good for this, because many castings can be made reliably and quickly. Many familiar objects, including car door-handles, locks and bathroom fittings are often made from these alloys.

Brass-making is still an important use of zinc. Brass has good resistance to corrosion and is a good conductor of electricity. It is used for plumbing and electrical purposes.

A large amount of zinc is made into sheets. These can be used for roofing and guttering because zinc corrodes slowly. The common torch or radio battery contains small sheets of zinc. The working of a zinc battery is explained on page 115.

The main areas in which zinc is used today are shown in *figure 7.16*.

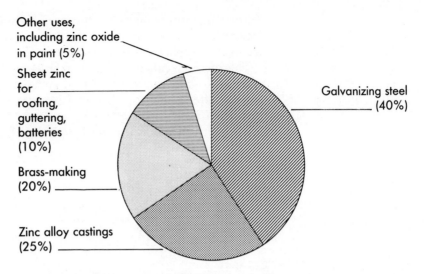

Figure 7.16 World use of zinc. Most zinc is used in the building industry.

Zinc blende—
the zinc ore

The common ore of zinc is called zinc blende. It is mostly zinc sulphide ZnS. Most zinc ore is found in the developed countries, including North America, Canada, Australia and several European countries. Two big discoveries of ore have recently been made, one in Canada and one in South Africa. They should provide us with zinc well into the next century, although this is not a very long time.

Zinc blende is usually mined underground. The ore is crushed, and the zinc sulphide is separated by a method called *flotation*. To do this, the crushed ore is mixed with water, and air is blown through it. Bubbles of air stick to the pieces of ore and carry them to the surface. Any impurities stay behind.

Extracting zinc

Although zinc is a reactive metal, it can be extracted by reduction with carbon (like iron, which is just below it in the activity series). However, it is often extracted by electrolysis (like aluminium, which is just above it in the activity series).

Whichever method is chosen, the zinc sulphide must first be turned into zinc oxide. This is done by heating the ore strongly in air:

Zinc sulphide + Oxygen → Zinc oxide + Sulphur dioxide

Sulphur dioxide gas is formed, but this is not wasted. It is turned into sulphuric acid, which can be sold.

The zinc oxide which is formed can be reduced in a blast furnace, similar to the one used for extracting iron from iron ore.

Zinc oxide + Carbon → Zinc + Carbon monoxide

Aluminium and Other Metals

At the temperature inside the furnace, the zinc is a gas. It can be condensed in a spray of molten lead. The mixture of zinc and lead is cooled to separate the zinc from it.

This method, called the Imperial Smelting Process, was invented in the U.K. to extract zinc and lead at the same time. Zinc and lead ores are often found together, which is why the method is so useful (*figure 7.17*).

Alternatively, the zinc oxide can be dissolved in sulphuric acid and electrolysed.

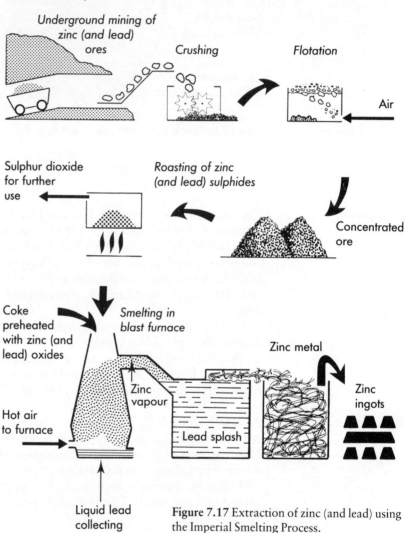

Figure 7.17 Extraction of zinc (and lead) using the Imperial Smelting Process.

Zinc in Batteries

Electricity can cause a chemical change, if it is passed through an electrolyte. This is called electrolysis, and it can be used to extract reactive metals like aluminium or zinc from their ores. It is possible for a chemical change to produce electricity. This is what happens inside a battery, or electric cell.

114 Chemistry in Use

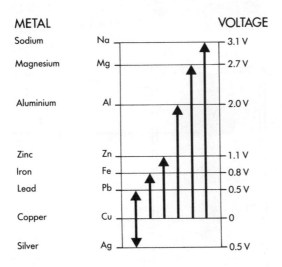

Figure 7.18 Voltages which can be made between copper and other metals.

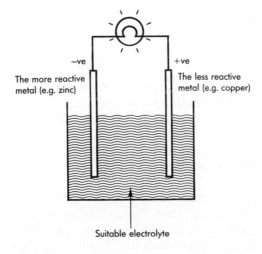

Figure 7.19 A simple cell, using two metals to make an electrical voltage.

Using chemicals to make electricity

When two different metals are placed in an electrolyte and connected by a wire, an electrical voltage is set up, and an electric current flows in the wire. In general, the further apart the two metals are in the activity series, the bigger the electrical voltage which is made. For example, zinc and copper can produce 1.1 V, while lead and copper would give only 0.5 V (*figure 7.18*).

The more reactive metal becomes the negative electrode of the battery, while the less reactive one becomes the positive electrode (*figure 7.19*).

The more reactive metal dissolves while electricity is being made. When it has all dissolved, there is no more chemical reaction and no more electricity is made.

Aluminium and Other Metals

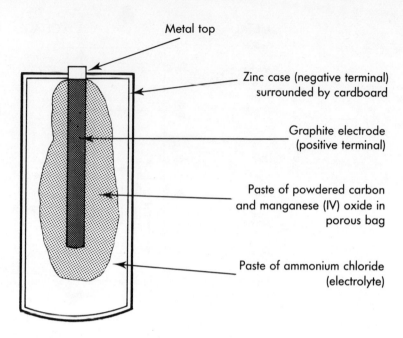

Figure 7.20 The primary cell or dry cell used in torch and radio batteries.

Primary cells

A primary cell is a battery which cannot be recharged. Once the chemical reaction is over, the battery is useless.

Of all the metals, zinc is most commonly used for this type of battery. The zinc is used as the negative electrode. In most primary cells, the positive electrode is not a metal. Instead, it is carbon in the form of graphite. A diagram of a primary cell is shown in *figure 7.20*.

A modern type of the primary cell has powdered zinc instead of a solid slab of zinc. This cell is more expensive, but more electrical power can be taken from it. Both the cells work in the same way. The zinc electrode slowly dissolves, until it eventually runs out.

The re-useable type of battery, called a secondary cell, is described on page 118.

Lead (Pb)

Lead has been known and used for a long time because it is quite easy to extract from its ores. The Romans used it to make water pipes and weights. Lead compounds have been used in paints for thousands of years. These paints have to be used carefully today because lead compounds are poisonous. It is now illegal in the U.K. to paint childrens' toys with lead paints.

Using lead

Lead is different from all the other common metals because it is soft and has little strength. This means that it has fewer uses than the harder and stronger metals like iron, aluminium, zinc and copper.

116 Chemistry in Use

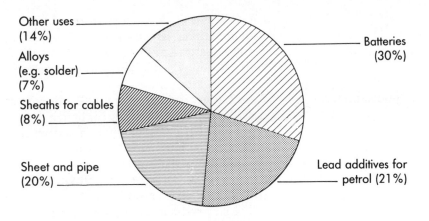

Figure 7.21 The uses of lead in the U.K. (1981).

The largest single use of lead is in storage batteries. Most vehicles on the roads use these batteries for starting and for running their electrical systems. Some vehicles, like milk floats, get all their energy from lead batteries instead of using petrol. Lead batteries are described on page 118.

In Britain and in many other countries, lead compounds are added to petrol to help the fuel to burn better. This causes problems of pollution (p.208).

Although lead is a soft and weak metal, it resists corrosion well. For this reason, lead sheet can be used for roofing, and lead pipes can be used for carrying water. Some lead from pipes does dissolve in water, especially in "soft" water areas. Since lead compounds are poisonous, lead pipes should not be used for carrying drinking water.

The main areas in which lead is used are shown in *figure 7.21*.

Alloys of lead

Lead melts at a low temperature for a metal. It can be mixed with other metals to make special alloys with even lower melting points. Solder is one example. Solder is an alloy of lead and tin, though it may contain other metals as well. Solder can be used to join two pieces of metal together. It is placed between the two pieces of metal and heated until it melts. When it cools down again it solidifies, joining the pieces of metal together. Wires in equipment like televisions, radios and electric motors are joined by soldering.

Many different solders are now available. They can even be used to solder materials like glass and ceramics. Copper water pipes are often joined by lead solder, which can cause pollution of drinking water.

One unusual alloy of lead finds use in fire sprinklers, which can be seen in many department stores, factories and offices. This alloy,

Aluminium and Other Metals 117

called Wood's metal, is a mixture of lead, bismuth, tin and cadmium. It has a melting point of only 70°C, less than the boiling point of water. The sprinkler is sealed with Wood's metal. If a fire starts, the alloy melts. This releases water to put out the fire.

Lead in batteries

A car battery is different from the zinc battery or primary cell described on page 116. The car battery has to be "charged" before it can be used. This is done by passing an electric current through it. While it is charging, it stores up electrical energy. This energy can later be used to start the car, or work the lights. A car battery like this is called a **secondary cell**, because it cannot be used straightaway like a primary cell.

The electrodes of car batteries are made of lead. These electrodes are dipped into an electrolyte of sulphuric acid. The reactions which give out electrical energy inside the lead battery are quite complicated. Sulphuric acid is used up when the battery is working, and the concentration of sulphuric acid can be measured to see if the battery needs recharging.

Batteries are a good way of using lead, because most of the metal can be used again after the battery has worn out. About 80% of battery lead is recycled in the U.K.

Figure 7.22 A British Rail Class 87 electric locomotive. Each of these engines contains 6 tons of copper in its electrical system.

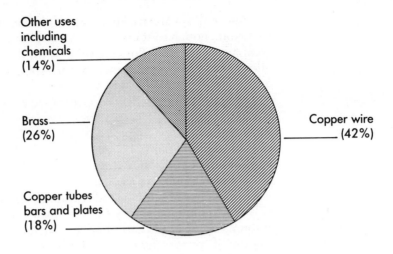

Figure 7.23 The uses of copper in the U.K. (1980).

Copper (Cu)

Using copper

Copper is mankind's oldest metal. It is possible that copper was used 10 000 years ago, while 8000 years ago the Egyptians were making copper knives and ornaments.

Copper is one of the best conductors of electricity after silver, which is much more expensive. The use of copper increased greatly after about 1850 when electricity was beginning to be harnessed. There are many other things about copper which make it useful for electrical wiring. It is soft and ductile, which means that it can easily be drawn into wires. It can be soldered easily and it does not corrode quickly.

Copper also conducts heat well. It is a good choice for making the bases of cooking pans. Heat is conducted quickly and evenly to all the food.

Copper is an unreactive metal, which is why it corrodes so slowly. It does not react with water, so it can be used as a roofing material or for plumbing. Hot and cold water pipes, including central heating pipes, are often made of copper.

Pure copper is quite a soft metal. Harder materials can be formed by making the copper into alloys.

Alloys of copper

The two main alloys of copper are bronze and brass.

Bronze is the alloy of copper (90%) with tin (10%). Large objects like ships' propellers are often made of bronze. Molten bronze is poured into a mould of the right shape. It then solidifies into the shape required. Bronze statues are made in the same way.

Brass is the alloy of copper (60%) with zinc (40%). About a quarter of the world's use of copper goes to making brass each year. There is more about brass on page 112.

Copper alloys are used to make coins. The metal for a coin has to be soft enough to stamp in the Mint, but hard enough not to wear away. Copper alloys are suitable for this. "Copper" coins are alloys of copper, tin and zinc. "Silver" coins are alloys of copper and nickel.

The main areas in which copper is used are shown in *figure 7.23*.

Ores of copper

The common ore of copper is copper sulphide. It is sometimes possible to find lumps of copper itself, called "native" copper, because copper is so unreactive. Native copper is quite rare, and it is not worth mining on its own.

Most copper ores mined today contain very little copper. It is only worth mining them by surface (open-cast) methods. Huge amounts of rocks and earth have to be removed to get a small amount of copper (*figure 7.24*). It is so expensive to fill in the mine again afterwards that this is not done. This means that large areas of countryside are destroyed. To make the Bougainville mine in the Solomon Islands, about 40 million tonnes of earth and forest had to be removed. 400 million tonnes of mined material were dumped in a nearby valley, choking water courses and silting the coastline.

Some developing countries, like Chile and Zaire, earn much of their money by selling copper ore. If the world price of copper goes down, they are in serious trouble. They have the same problem as some countries which produce aluminium ore. The U.K. used to supply over half the world's copper ore. Ores in the U.K. have now run out. Countries like America, Canada, Russia, Chile and Zaire now supply most of the copper.

Some people calculate that copper will run out within our lifetimes. Possibly this will not happen, because new ores may be discovered. Even if copper does not run out so soon, it will probably become more expensive, and there will be more destruction of the environment, as people search for poorer and poorer ores.

Extracting copper

Just like zinc ore, the copper ore is first ground up and then separated by flotation (*figure 7.25*). The purified copper ore is then roasted in air. This burns off the sulphur in the copper sulphide. It also burns off impurities like iron. The copper which is left behind is called blister copper, because it looks as though it has blisters all over it. Blister copper is about 98% pure copper. This is not good enough for most uses, and the copper is purified by electrolysis.

Getting pure copper by electrolysis

Large slabs of blister copper, each weighing about 300 kg, are dipped into a solution of copper sulphate. The slabs of copper are connected to the positive terminal of an electricity supply. This means that the blister copper itself becomes the positive electrode. The negative electrodes are thin sheets of pure copper (*figure 7.26*).

Figure 7.24 An open-cast copper mine in Papua New Guinea. More careful recycling and use of copper could save areas of countryside like this.

Figure 7.25 Inside the control room of a copper mine. This section is used to control the crusher which grinds up the copper ore.

Figure 7.26 Purifying copper by electrolysis. A batch of negative electrodes (cathodes) is lifted by a crane.

When the current is switched on, the blister copper slowly dissolves. At the same time, more pure copper is formed on the copper sheets. What is happening is that copper from the slabs is transferred to the pure copper sheets. The pure copper is taken out, melted down, and made into wires or bars.

Any impurities in the blister copper fall to the bottom of the electrolysis tank. These include metals like gold, silver and platinum, which are valuable (*figure 7.27*).

Electroplating of Copper and Other Metals

The purification of copper by electrolysis is an example of electroplating. **Electroplating** means forming a layer of metal on something by using electrolysis.

The substance to be electroplated is made the negative electrode. When copper is purified by electroplating, the negative electrode starts as a thin slab of pure copper. More pure copper is plated on top of it. The negative electrode is dipped into a solution containing a compound of the metal required for plating. For example, if copper plating is wanted, a solution of copper sulphate is suitable. Copper sulphate solution contains positively charged copper ions. During electrolysis they go to the negative electrode. At the negative electrode they gain electrons and are turned into copper atoms. These copper atoms cover the electrode (*figure 7.29*).

122 Chemistry in Use

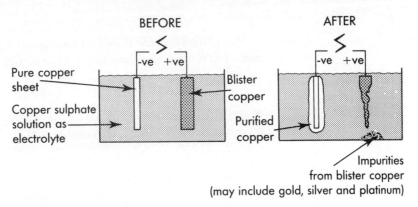

Figure 7.27 Purifying copper by electrolysis.

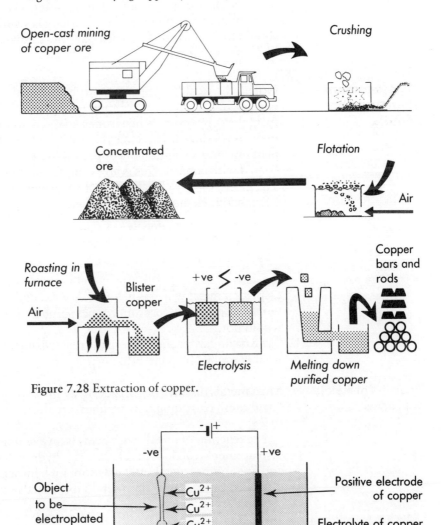

Figure 7.28 Extraction of copper.

Figure 7.29 Electroplating, using copper. The same idea can be used for chromium plating and nickel plating.

Aluminium and Other Metals 123

This can be described by a simple equation. (Compare it with the equation for making aluminium on page 109.)

Copper ions + Electrons → Copper
$Cu^{2+} + 2e^- \quad\quad Cu$

When blister copper is purified by this method, the blister copper dissolves in the solution. This means that any copper ions from the copper sulphate which are turned into copper are immediately replaced.

A similar method can be used for chromium plating or silver plating. Steel car bumpers are often chromium-plated to prevent rust and to make them look shiny. The steel bumper is made into a negative electrode. It is dipped into a solution containing a chromium compound and the electricity is switched on. You should by now be able to work out how to silver-plate something like a nickel spoon.

Displacement Reactions of Metals

Not all the world's copper is made by the method already described. Sometimes, a solution of copper sulphate is first made from the copper ore. When iron is mixed with this solution, copper metal is produced. The same thing will happen if you dip a steel penknife into a solution of copper sulphate. The knife becomes coated with copper. This is an example of a **displacement reaction**. The more reactive element, iron, has displaced or pushed out the less reactive element, copper:

Iron + Copper sulphate → Copper + Iron sulphate
$Fe + CuSO_4 \quad\quad\quad Cu + FeSO_4$

You can tell if a displacement reaction would work by looking at the activity series (p.79). Any metal in the activity series will displace a metal which is below it. For example, magnesium or zinc would displace copper from copper sulphate solution, but silver would not.

Two similar reactions using the reactive metal aluminium have been described on page 110.

Reaction of metals with acids

Most metals dissolve or corrode away in acids. The problem of "acid rain" corroding railway lines has already been mentioned (p.23).

The reaction of metals with acids is an example of a displacement reaction. In this case it is hydrogen which is displaced and not another metal. All acids contain hydrogen, which can be displaced by lead or by any metal more reactive than lead. For example:

Zinc + Hydrochloric acid → Zinc chloride + Hydrogen
$Zn + 2HCl \quad\quad\quad\quad ZnCl_2 + H_2$

Magnesium + Sulphuric acid → Magnesium sulphate + Hydrogen
$Mg + H_2SO_4 \quad\quad\quad\quad\quad MgSO_4 + H_2$

Copper and the less reactive metals do not normally react with acids.

The Most Reactive Metals

Potassium, sodium and magnesium are three of the most reactive metals. The metals themselves do not have many uses, because they corrode so easily. All of them react with air and water, which is always present in the atmosphere.

These metals are more reactive than aluminium, so they have to be extracted from their ores by electrolysis. The ores are usually the metal chlorides. Sodium chloride, common salt, is the best known example. The next chapter is all about the chemicals which are made from salt.

Potassium (K) and sodium (Na)

Potassium and sodium are so reactive that they cannot be kept in the open air. They have to be stored under oil. You may have already seen what happens when small pieces of potassium or sodium are dropped into some water. They rush about on the surface, producing bubbles of hydrogen gas as they react. An alkali, potassium or sodium hydroxide, is also made in this reaction. For this reason, potassium and sodium are sometimes called the *alkali metals*:

Sodium + Water → Sodium hydroxide + Hydrogen
$2Na + 2H_2O \qquad 2NaOH + H_2$

Potassium + Water → Potassium hydroxide + Hydrogen
$2K + 2H_2O \qquad 2KOH + H_2$

Potassium and sodium conduct electricity and heat well, like all metals, but they are different in other ways. They are so soft that they can be cut by a knife, and they are so light that they float on water. They also melt very easily. The low melting point of sodium and its good conduction of heat make liquid sodium ideal for cooling the fast-breeder type of nuclear power station. The most familiar use of sodium is in street lighting, because it can be made to give out an orange glow.

Magnesium (Mg)

Magnesium is less reactive than potassium or sodium, but it still corrodes easily.

The reactivity of magnesium is seen when it burns in air. It reacts with oxygen in the air, giving out a bright white light. The equation for this reaction is:

Magnesium + Oxygen → Magnesium oxide
$2Mg + O_2 \qquad 2MgO$

Flares, fireworks and some flash bulbs include magnesium to produce this bright light.

Magnesium is a light metal. It does not have great strength on its own, but strong alloys can be made from it. The lightness of these alloys makes them especially useful (*figure 7.30*).

Figure 7.30 The main landing wheel of an F27 aircraft, made from a magnesium alloy. The lightness and strength of magnesium alloys make them ideal choices for this type of use.

Questions

1. Complete the following table about electron arrangements. You will need to use the Periodic Table on p.142.

Particle	Symbol	Atomic number of element	Number of electrons	Electron arrangement
Magnesium atom	Mg	12	12	2,8,2
Chlorine atom	Cl			
Chloride ion	Cl^-			
Calcium ion	Ca^{2+}			
Sulphur atom	S			
Oxide ion	O^{2-}			
Aluminium ion	Al^{3+}			
Fluoride ion	F^-			
Nitrogen atom	N			
Potassium atom	K			
Sodium ion	Na^+			

2. Titanium is made by reacting titanium (IV) chloride $TiCl_4$ with magnesium.
   (a) Write word and symbol equations for this reaction.
   (b) What does this tell you about the position of titanium in the activity series?

3. Copper is used for making water pipes, electrical wires and coins. New supplies of copper are becoming hard to find.
   (a) Which properties of copper make it a good choice for plumbing?
   (b) Explain what you might use to replace copper for piping if copper became difficult to buy
   (c) Cables made of glass fibre instead of copper are now used to carry information from one place to another. More information can be carried using glass rather than copper. In addition, copper can be saved for other purposes. Could all copper wires be replaced by glass? If not, what could be used instead and what would be the problems?
   (d) What will happen to "copper" coins if the price of copper increases greatly?

4. Find the current prices of some metals from a newspaper. Suggest reasons for some of the differences which you find.

5. Explain, with the aid of a labelled diagram, how you would plate a piece of steel with nickel using electroplating.

8 Chemicals from Salt

Salt (sodium chloride) has always been an important chemical. In earlier times it was essential for preserving meat and other foods. It was also needed for tanning leather. For both these reasons, salt was vital to armies. Armies need feeding and they needed leather for armour and harness. Salt was so important to the Romans that they even paid their soldiers with it. The word "salary" comes from the Latin word "sal", which means salt.

Today we have other ways of keeping food and we do not use leather for armour. Instead, salt has become important to us in many other ways. We use it to keep roads clear of snow and ice in the winter. Chemicals made from salt are needed for making PVC plastics, bleaches, glass, soap and textiles, to name only a few. Enough salt is mined in the U.K. each year to give each person over 100 kg.

Extracting Salt

Salt can be found underground, like other minerals. There are several salt mines in the U.K. Most of these are in Cheshire, Lancashire and Yorkshire. These salt deposits were formed by the evaporation of an ancient sea about 180 million years ago. Solid rock salt is mined in only one British mine, the Winsford mine in Cheshire (figure 8.1). The salt from this mine is bought by councils to put on the roads.

In all the other mines, salt is brought to the surface by pumping water down. The salt dissolves in the water, leaving most other rocks and minerals behind. The salt solution, which is called brine, is pumped back to the surface. The brine itself can be used as a raw material for making other chemicals.

Every litre of sea water contains about 25 g of salt. In hot countries, the salt can be obtained by using the sun to evaporate the water. This leaves the salt behind. The sea water is collected in huge ponds, which are often dyed green to trap more of the solar energy. In one place in Australia over a million tonnes of water are evaporated each day (figure 8.2). The same method is used, on a smaller scale, in many countries next to the Pacific and Indian oceans.

Figure 8.1 The Winsford salt mine in Cheshire, U.K. Explosives are used to blast away the rock salt. Huge trucks are then used to carry the rock salt through the mine.

Figure 8.2 Extracting sea salt at Dampier, Australia. Sea water is pumped slowly from pond to pond. As the water evaporates in the heat of the sun, the salt solution becomes more concentrated. After about eighteen months, the salt solution is concentrated enough to crystallize. The crystallizing ponds can be seen clearly in the photo. On the surface of these ponds can be seen the 50-tonne trailers which carry the salt to the stockpiles in the background.

Using Salt and Brine

Salt on the roads

More salt is used on the roads in the U.K. than for any other purpose. Water normally freezes at 0°C, so the roads are often covered by ice in the winter. If salt is added, the mixture of salt and water freezes below 0°C. This means that ice does not form on the roads unless the temperature is much lower than 0°C. Roads may be clear of ice if the air temperature is as low as −12°C.

Sodium from salt

Sodium is a very reactive metal. Like other reactive metals, including aluminium, it has to be made by electrolysis. Sodium chloride is used for this. The salt is melted and then electrolysed.

Sodium is so reactive that there are not many uses for it (p.125). Only a few hundred thousand tonnes of it are made each year, compared with millions of tonnes of some other metals.

Chemicals from brine

Salt, in the form of brine, is a vital raw material for industry today. Chlorine, sodium hydroxide and sodium carbonate are the three most important chemicals which can be made from it. Two of these chemicals, chlorine and sodium hydroxide, can be made by the electrolysis of brine. The third, sodium carbonate, is made from brine by a series of chemical reactions.

Figure 8.3 The major use of salt in the U.K. is for keeping roads clear of ice. In severe conditions, 1 tonne of salt may be spread on every 5 km of a main road.

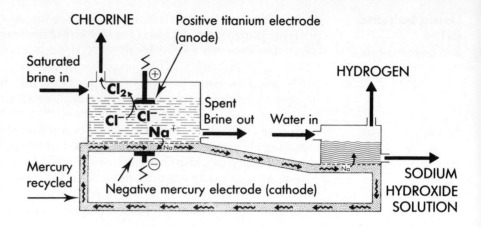

Figure 8.4 The Castner-Kellner cell for the electrolysis of brine. The three products are chlorine, sodium hydroxide solution and hydrogen.

The Castner-Kellner process

The manufacture of chlorine and sodium hydroxide together is known as the chlor-alkali industry. There are several methods used today for the electrolysis of brine. The most common of these in the U.K. is the Castner-Kellner process.

A diagram of this process is shown in *figure 8.4*. Keep referring to it while you read how the method works.

The electrolysis cell, known as a mercury cell, is on the left of the diagram. It is called a mercury cell because the negative electrode is liquid mercury. The positive electrode is made of titanium. The electrolysis cell is filled with brine.

Sodium chloride is made of positive sodium ions and negative chloride ions. The negative chloride ions are attracted to the positive electrode during electrolysis. When they reach this electrode, they are turned into chlorine molecules by giving up their electrons (negative charges):

Chloride ions − Electrons → Chlorine molecules
$2Cl^- - 2e^-$ Cl_2

The chlorine gas leaves the cell. It is collected and stored for use. The positive sodium ions are attracted to the negative mercury electrode. At this electrode they gain electrons to become atoms of sodium metal:

Sodium ion + Electron → Sodium atom
$Na^+ + e^-$ Na

The sodium dissolves in the mercury. It is carried away from the electrolysis cell as the mercury is pumped round.

130 Chemistry in Use

The mixture of the metals mercury and sodium is an example of an alloy. Alloys of mercury are called **amalgams**. If you have a filling in a tooth, your dentist will probably use an amalgam of mercury with silver and tin.

The amalgam of mercury and sodium is pumped round to a chamber containing pure water. The sodium reacts with the water. This makes a solution of sodium hydroxide, together with hydrogen gas. You may have seen the same reaction when a piece of sodium is dropped into water.

Sodium + Water → Sodium hydroxide + Hydrogen
$2Na + 2H_2O \qquad\qquad 2NaOH + H_2$

The sodium hydroxide and the hydrogen are collected, while the mercury is sent round again.

Problems with mercury cells

In an ideal world, no mercury would be lost from a mercury cell. In practice there is always a leakage of mercury. Mercury is one of the "heavy" metals, like lead, which can cause severe poisoning. It is known to cause brain damage, just like lead. The expression "mad as a hatter" may have started because hatters used mercury compounds. There is a theory that Napoleon went mad because of mercury poisoning.

During the 1950s there was a tragedy at a place called Minamata in Japan. It was caused by mercury pollution. Mercury leaked into the sea and found its way through the food chain into fish. The fish were then eaten by local people. Over 40 people died of the poisoning, and more than 60 were disabled for life.

Because of the problem of mercury poisoning, no new mercury cells are likely to be built in the major industrial countries. Scientists are now working on a new method for the electrolysis of brine which does not need mercury.

Chlorine

Chlorine is one of the *halogen* family of elements. It is difficult to store safely because it is a reactive and poisonous gas. The horror of chemical warfare started with this dense, green, poisonous gas. At 5 p.m. on the afternoon of 22nd April, 1915, during the First World War, the Germans released 180 tonnes of chlorine gas at Ypres. It was carried by the wind and rolled into the French trenches. Within 2 hours, 5000 men were dead or dying. Both sides, the Allies and the Germans, used poisonous gases until the end of the war. The gases killed over 100 000 people and affected a million more. Even worse chemical weapons are made and stockpiled today (p.214).

Chlorine is used today in a wide range of chemicals. The importance of chlorine for different purposes is shown in *figure 8.5*.

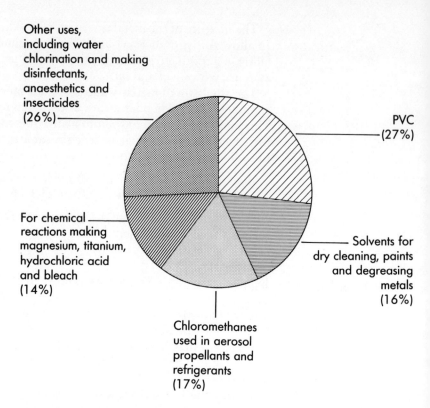

Figure 8.5 The uses of chlorine in the U.K.

PVC plastic

PVC, which stands for poly(vinylchloride), is a common plastic. The making of PVC and its importance are described on page 68. More chlorine is used to make PVC than for any other single use. As our demand for PVC has risen, so has the amount of chlorine made (*figure 8.6*).

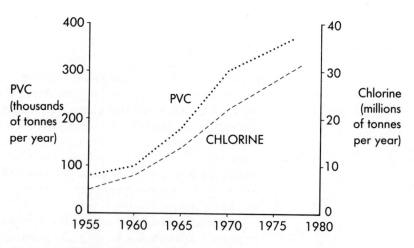

Figure 8.6 The world's increasing demand for PVC and chlorine.

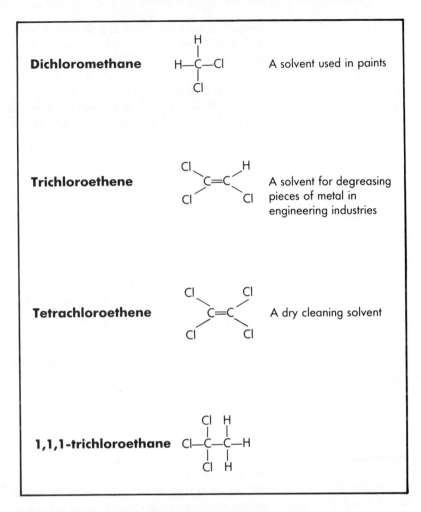

Figure 8.7 Some organic solvents which are made from chlorine. Notice how the chemicals are named. Dichloromethane is basically methane with two chlorine atoms (*di*chloro-). Trichloroethene is basically ethene with three chlorine atoms (*tri*chloro-). "Tetrachloro-" means four chlorine atoms.

1,1,1-trichloroethane has this name because all three chlorine atoms are joined to the first carbon atom.

Solvents

A **solvent** is a chemical which is good at dissolving other chemicals. Water is a good solvent for many chemicals (p.184). However, it is not good at dissolving organic chemicals (compounds of carbon) like greases, oils and most glues and paints.

An organic solvent is usually needed to dissolve these organic chemicals. Many useful organic solvents contain chlorine. The formulas of some of these solvents are shown in *figure 8.7*. They include solvents used in paints, for degreasing metal parts of cars and machines and for dry-cleaning.

The word "dry-cleaning" means cleaning using organic liquids. Dry-cleaning solvents like tetrachloroethene are much better than water at removing oily and greasy dirt. Water is usually added as well as the dry cleaner, so that other types of dirt are washed out as well.

A bottle of Tipp-Ex contains a typical organic solvent. The white chemical which is used to cover any mistakes is dissolved in the solvent. When the solution is put over the mistake, the solvent evaporates. This leaves the white chemical behind. You should find the name of the solvent, 1,1,1-trichloroethane, on the label of a Tipp-Ex bottle.

Bleaches, disinfectants and other killer chemicals

A familiar test for chlorine gas is the bleaching of moist litmus paper. Compounds of chlorine are found in many household bleaches. The chemistry of bleaches is explained on page 202.

Many people recognise chlorine from the smell in swimming pools. Chlorine is used as a disinfectant in swimming pools and in ordinary water supplies because it kills many germs.

Chemicals made from chlorine are also used as pesticides (pest killers) to help farmers. Some of these compounds, including DDT and 2,4,5-T, are described on page 172.

Only a small amount of the world's chlorine is used for making bleaches and disinfectants. Even so, these are important uses, especially for the developing countries. A clean water supply is essential for health.

Sodium Hydroxide

Sodium hydroxide was first used in large amounts for making soap. It is still used for making soaps and detergents today, but it now has many other uses.

Sodium hydroxide is a cheap alkali, so it can be used to neutralise unwanted acids (p.162). Strong alkalis like sodium hydroxide are also good at reacting with and dissolving many oils and greases. Indeed, soap is made from fats using this reaction (p.199). Sodium hydroxide is used in powerful oven cleaners (p.202) and in the cleaning of textiles as they are being made.

Some of the many uses of sodium hydroxide today can be seen by looking at *figure 8.8*.

Hydrogen

Large amounts of hydrogen are made during the electrolysis of brine. Hydrogen is difficult to handle because it is a gas and because it burns easily. It can explode when it is mixed with air. Every time you test for hydrogen gas in the laboratory, using a lighted splint, you are trying to make it explode in air.

Hydrogen has such a low boiling point that it is difficult to liquefy. It is usually carried as a gas in heavy, high-pressure cylinders. This makes it expensive to move, because one lorry cannot carry much gas (*figure 8.9*).

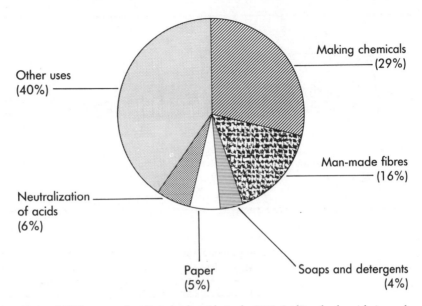

Figure 8.8 The uses of sodium hydroxide in the U.K. Sodium hydroxide is used in a vast number of different chemical reactions.

Figure 8.9 Transport of hydrogen. A trailer carrying hydrogen gas may weigh about 30 000 kg. The gas itself will weigh only about 100 kg.

Making margarine

Pure hydrogen is needed for many reactions involving organic chemicals. One example is the hardening of plant oils to make margarines. Plant oils contain molecules with double bonds between carbon atoms, like alkenes (p.66). Molecules like this have low melting points and are often liquid at room temperature. If the molecules are reacted with hydrogen, the double bonds become ordinary single bonds, as in alkanes. These new molecules have higher melting points. They are often solid at room

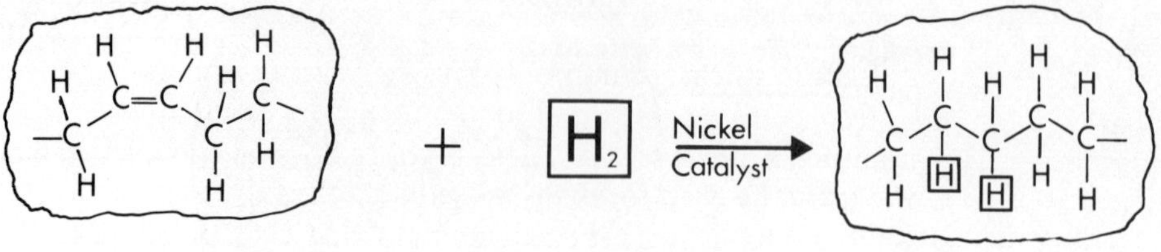

Part of a plant oil molecule

Part of a margarine molecule

Figure 8.10 Making a *hard* margarine. Plant oils are reacted with hydrogen, using a nickel catalyst, to turn them into harder margarines.

Molecules which contain double bonds between carbon atoms, like the plant oils, are called "unsaturated". Molecules with no double bonds are called "saturated". If the oils contain many double bonds, they are known as "polyunsaturated" oils. When you are next in a grocery, see if you can find a margarine which contains polyunsaturated oils. It should be quite runny and easy to spread.

temperature (*figure 8.10*).

The margarine can be made as soft or as hard as needed by using the right amount of hydrogen. Margarine is cheaper than butter because it comes from plant oils and not from animals.

Burning hydrogen

Hydrogen burns so well that the early space rockets burnt a mixture of hydrogen and oxygen. The hot flame from burning hydrogen can also be used for welding metals together.

Any hydrogen which is made is never wasted. If it cannot be sold, it can be burnt inside the works to provide useful energy, or it can be reacted with chlorine to make hydrochloric acid.

If a cheap way can be found of making hydrogen from water, using solar power, it could even be the fuel of the future (p.52).

Glass

Salt is an important chemical in glass making. The three main raw materials for glass are sand (silicon dioxide), limestone (calcium carbonate) and sodium carbonate.

The sand and limestone can be quarried. The sodium carbonate is made from salt and limestone. About half the sodium carbonate in the U.K. is produced for making glass.

Making glass

Glass is made by melting sand with other chemicals. Sand melts at a very high temperature and the sodium carbonate is added to lower its melting point. The sand and sodium carbonate react to make sodium silicate:

Silicon dioxide + Sodium carbonate → Sodium silicate + Carbon dioxide
(sand)
$SiO_2 + Na_2CO_3$ $Na_2SiO_3 + CO_2$

Sodium silicate, commonly known as water-glass, dissolves in water. Limestone is added so that a mixture of calcium silicate and sodium silicate is formed. This mixture does not dissolve in water, and it can be used for making ordinary glass.

Other chemicals, especially metal oxides, can be added to give different sorts of glass. Chromium oxide is used to make green glass. Blue glass can be made using cobalt oxide.

The amounts of these chemicals which are used for making glass are shown in *figure 8.11*.

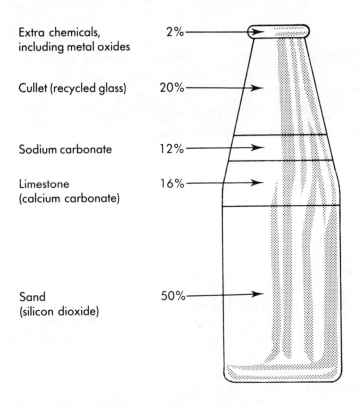

Figure 8.11 Chemicals used in glass making.

Shaping glass

Think of all the different shapes of glass needed to make bottles, jars and window panes. Glass can be shaped as the hot liquid cools. Bottles, jars and similar objects can be made if the glass is allowed to cool in moulds.

Flat glass for windows is made by letting the glass cool on a bath of molten tin. The surface of the liquid tin is smooth, so the glass becomes smooth as well. Pieces of glass can then be cut off for customers as required (*figure 8.12*).

Figure 8.12 A continuous sheet of glass made by the float glass process.

Glass is not really a solid, although it looks like one. Glass from windows a few hundred years old is thicker at the bottom than the top. The glass has flowed downwards like a liquid.

Even though glass is a liquid, it can be made strong for use in different types of safety glass. If glass is heated and then cooled quickly, toughened glass is made. Toughened glass can be five times stronger than ordinary glass. This sort of glass is useful in car windscreens, because it shatters into small pieces when broken, instead of giving dangerous splinters.

Waste glass

Six billion glass bottles and jars were made in the U.K. in 1980. Only 15% of these were designed to be returnable. Two million tonnes of glass are thrown away each year, making up 10% of all household refuse. Apart from the waste, glass which is just thrown down as litter is dangerous. It can cause injuries to animals or people who step on it, and it can cause road accidents. Glass may even cause fires, because it can focus the sun's rays like a magnifying glass.

Waste glass is known as cullet. It is a good idea to recycle this cullet as much as possible. Some cullet is already used in making new glass. Not only does it save raw materials like sand and limestone, but it also saves energy. This is because cullet makes the mixture used for making glass melt at a lower temperature. Less energy is therefore used in glass making, which saves fossil fuels and money.

Only 20% of cullet is generally used in glass making. It is possible to use much higher amounts, if the cullet is available. The glass industry has started a scheme for collecting more cullet, using Bottle Banks (*figure 8.13*). Where is your nearest Bottle Bank? If it is near you, do you use it?

The Bottle Bank scheme is operated by local authorities in conjunction with glass container manufacturers, under the co-ordination of the Glass Manufacturers Federation, the industry's trade association.

Bottle Banks are large, specially-designed skips which hold around three tonnes of waste glass. They are placed in locations normally visited by large numbers of people in their day-to-day or weekly business, such as supermarket or town centre car parks.

Members of the public are asked to deposit their waste bottles and jars in the Bottle Banks in separate compartments for clear, brown or green glass. (In some areas Bottle Banks may have only two compartments for clear and mixed brown and green glass.)

The local Council empties the containers at regular intervals and sells the waste glass — cullet as it is known in the glass industry — to a glass manufacturer who mixes it with other raw materials and remelts it to make new glass containers.

Glass Manufacturers offer local authorities three important safeguards.

>A guaranteed price for the cullet
>A guaranteed market
>An assurance that local authorities will not be encouraged to set up a scheme unless its operation would at least break even.

Bottle Bank: the Scheme Explained

Bottle Bank: The Benefits

National energy savings

Better use of natural resources

Saving in refuse disposal costs

Extra source of income for local authorities

Public education of the need to save waste

Figure 8.13 The Bottle Bank Scheme.

Questions

1. The metal sodium is extracted from sodium chloride by electrolysis. The method is known as the Downs Process:

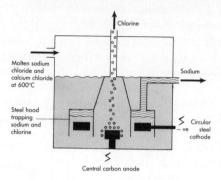

(a) Why is sodium extracted by electrolysis?
(b) Explain why sodium is made at the negative electrode. Write an equation for the reaction which takes place at this electrode.
(c) Explain why chlorine is made at the negative electrode and write the equation for the reaction which takes place at this electrode.
(d) Suggest a reason why calcium chloride is added to the sodium chloride.
(e) Explain whether or not you expect the sodium to be pure.
(f) Why is the positive electrode not made of steel?

2. Oxidation can be defined as addition of oxygen, while reduction is removal of oxygen. This is not the only way of looking at oxidation and reduction (redox) reactions. Oxidation can be defined as loss of electrons. Substances which lose electrons are oxidized. Reduction can be defined as gain of electrons. Substances which gain electrons are reduced.

Using this idea, a substance which removes electrons from another substance is an oxidizing agent. A reducing agent gives away electrons.

(a) When sodium chloride is electrolysed in the Castner-Kellner process, the reaction at the negative electrode is

$$Na^+ + e^- \rightarrow Na$$

(i) Is the sodium ion (Na^+) oxidized or reduced? Explain.
(ii) Is the negative electrode a reducing or an oxidizing agent?

(b) Chlorine reacts with sodium to form sodium chloride:

$$2Na + Cl_2 \rightarrow 2NaCl$$

(i) Is the sodium (Na) oxidized or reduced?
(ii) Is chlorine an oxidizing agent or a reducing agent?

3. Explain, step by step, how you would obtain pure salt from rock salt. Rock salt is a mixture of salt and sand.

4. What is a dry-cleaning solvent and why does it have this name?

9 Explaining the Properties of Chemicals

Why are diamonds so hard, making them useful for cutting and drilling? Why is graphite so smooth, so that it can be used for lubricating moving parts in motors? Why does helium not react with other chemicals, making it safe to use in airships? Why is carbon dioxide a gas? Why is the alloy of copper and nickel harder than pure copper, making it more suitable for long-lasting coins?

The properties of these and other chemicals can be explained by looking more closely at the particles which make up the chemicals themselves. Information about atoms, the smallest particles of the elements, is contained in the Periodic Table.

The Periodic Table

There are more than 100 elements. The Periodic Table is a way of organizing or classifying the elements so that it is easier to see patterns among them. Some of the ways of using the Periodic Table have been described in earlier chapters. One example is the difference between metals and non-metals. The metals are on the left of the Periodic Table, whereas the non-metals are on the right.

There is much more to the Periodic Table than the difference between metals and non-metals. Elements in the Periodic Table are divided into rows and columns.

A row of elements is called a **period**.

A column of elements is called a **group**.

Periods of elements

Elements are placed in the Periodic Table according to their atomic number. The atomic number of an element tells you how many protons and how many electrons there are in every atom of that element (p.33). In chemistry it is the number of electrons which matters most. The reason for this is that electrons are important in chemical reactions. The electrons are arranged round the nucleus of the atom in layers called electron shells. The arrangement of electrons in atoms is explained in detail on page 59. You should read through that section if you have not already done so. Each new shell of electrons is represented by a new period. There are seven electron shells, corresponding to the seven periods in the Periodic Table.

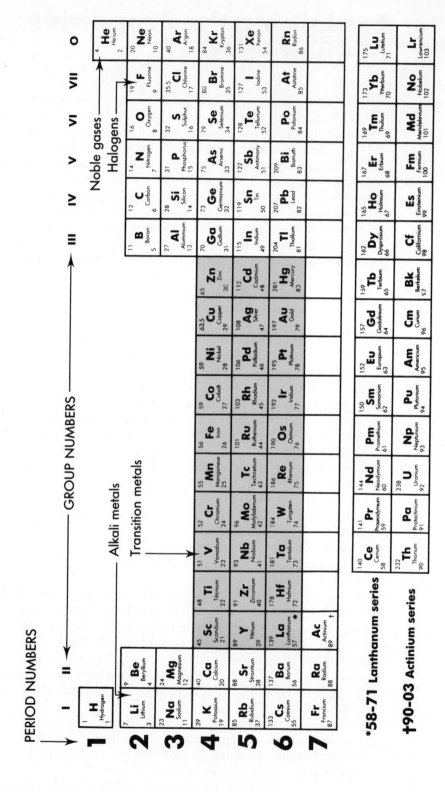

Figure 9.1 The Periodic Table.

Groups of elements

Elements in the same group have the same sorts of properties and chemical reactions. This can be explained by the arrangement of electrons in their atoms. Atoms of elements in the same group have the same number of electrons in their outermost electron shell. For example, atoms of the elements in Group I (the alkali metals) have one electron in their outermost shell. Atoms of the elements in Group VII (the halogens) have seven electrons in their outermost shell. Atoms with a similar arrangement of electrons have similar chemical reactions.

The noble gases, sometimes called Group O, form a special family of elements. This group, including helium, neon and argon, is the right-hand group in the Periodic Table. The noble gases hardly react with anything. Neon and argon are used to fill light bulbs and tubes because they are so unreactive. They will not react with the filament and burn it away, even at high temperatures. The atoms of the noble gases have full shells of electrons. This arrangement is stable, so that these elements do not easily react with anything else. This explains why a gas like helium can be used safely in airships. There is no chance of it catching fire and causing a disaster like the older hydrogen-filled airships.

Structure and Bonding in Chemicals

All chemicals are made of particles, which may be atoms, ions or molecules. The properties of a chemical depend on two things. One is the way in which the particles are built up together—this is called the **structure** of the chemical. The other is the type of attraction which holds the particles together—this is called the **bonding** in the chemical.

Metals and alloys

Lumps of metal are built up from metal atoms. The atoms are packed together closely in a regular way. A piece of metal is really, therefore, a crystal of metal atoms.

Most metals are hard solids with high melting points and high boiling points. This is because the metal atoms are held strongly to each other. The **metallic bonding**, shown in *figure 9.2*, holds all the metal atoms together in a big, strong structure.

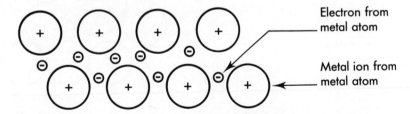

Figure 9.2 Metallic bonding. The outer electrons in metal atoms are free to move between the atoms. You can picture a piece of metal as layers of metal ions with electrons among them. Opposite electric charges attract each other, so the negative electrons hold the giant structure of positive metal ions together.

Explaining the Properties of Chemicals

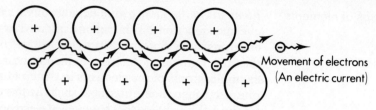

Figure 9.3 Conduction of electricity in metals. Electrons are so small that they can move among the metal atoms. Imagine a "sea of electrons" flowing through the metal and making up the electric current. Metals conduct heat in a similar way. Heat energy is passed on by the movement of electrons.

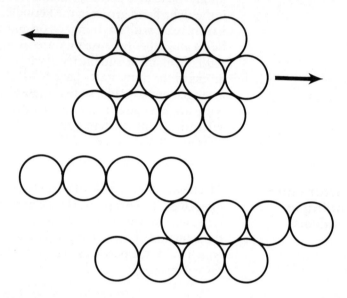

In the pure metal, layers of metal atoms can be made to slide over each other. The metal is malleable.

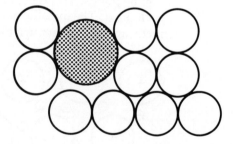

In the alloy, it is not so easy to move the metal atoms. Alloys are usually harder and less malleable and ductile than pure metals. They also generally have lower melting points and do not conduct electricity as well as the pure metals.

Figure 9.4 Pure metals and alloys.

Metals are the only common substances, apart from graphite and silicon, which conduct electricity when they are solid. This can also be explained by their structure. Electricity is conducted whenever electric charges move. The "sea of electrons" (negative electric charges) can move through the metal to conduct the electricity. Electrons can flow through metals, graphite and silicon, but not normally through anything else (*figure 9.3*).

Alloys are mixtures of different metals. They can be made by melting metals together. Alloys are usually harder and less malleable, i.e. less easy to work, than pure metals (*figure 9.4*). This is why "copper" coins are made of an alloy. They resist wear for longer than pure copper, so they stay in circulation for a longer time. Alloys also have lower melting points than pure metals. Solder is an example of a useful alloy with a low melting point (p.117).

Ionic substances

Ionic substances are built up from ions. Ions are atoms or groups of atoms which are electrically charged (p.104).

Salt (sodium chloride) is a typical ionic substance. It contains positively charged sodium ions (Na^+) and negatively charged chloride ions (Cl^-). Salt is a solid with a high melting point and high boiling point, because the sodium ions and chloride ions are held strongly together. The ions are held together because opposite electric charges attract each other. This is called **ionic bonding**. The sodium ions and chloride ions are built into a large, regular structure, held together by ionic bonding (*figure 9.5*).

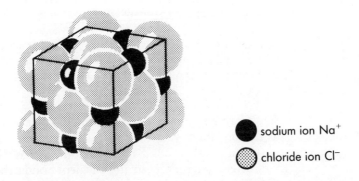

● sodium ion Na^+

◍ chloride ion Cl^-

Figure 9.5 Ionic bonding in sodium chloride. The ions are held together in a giant structure called a lattice.

Ionic substances do not conduct electricity when they are solid, because the ions cannot move to carry the current (p.104). Ionic substances will only conduct electricity if they are molten or dissolved in a liquid, when the ions are free to move. Ionic substances are electrolytes.

Explaining the Properties of Chemicals

Covalent substances

Covalent substances are made of molecules. The atoms which make up the molecules are held together by **covalent bonding** (p.60). Covalent substances are non-electrolytes.

Diamond and carbon dioxide are both covalent substances. The reason why they have such different properties is that diamonds are giant molecules, whereas carbon dioxide molecules are small.

Diamond is the hardest known natural substance. It is a form of pure carbon. The carbon atoms are joined together by covalent bonds into a giant structure. The covalent bonds are strong, so the structure is difficult to break down. This explains the hardness of diamond and its high melting and boiling points. Graphite is another form of pure carbon. These two forms of carbon are compared in *figure 9.6*.

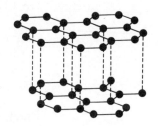

Diamond
The hardest known natural substance.
A non-conductor of electricity.
Density of 3.5g cm^{-3}.
Colourless and transparent crystals.

Graphite
A soft, flaky and slippery substance.
A conductor of electricity.
Density of 2.2g cm^{-3}.
Black and shiny.

Figure 9.6 Covalent bonding in diamond and graphite. Both these substances are found as giant structures. Different forms of the same element, like diamond and graphite, are known as allotropes.

Carbon dioxide is made of small molecules (*figure 9.8*). The carbon and oxygen atoms in each molecule are held together by strong covalent bonds. Although the atoms within each molecule are held together by strong covalent bonds, the molecules themselves are not much attracted to each other. This makes it easy to separate one molecule from another, so carbon dioxide becomes a gas at a low temperature. This explains why chemicals made of small molecules, like carbon dioxide, are usually gases or liquids at room temperature.

Figure 9.8 Covalent bonding in carbon dioxide. Each carbon dioxide molecule is held together by strong double bonds between the carbon and oxygen atoms. The molecules themselves have little attraction for each other. Separate carbon dioxide molecules form a gas at room temperature. ▶

Figure 9.7a Diamond in action. A saw with a diamond blade quickly cuts a concrete slab. Water is used to cool the diamond blade during the cutting operation.

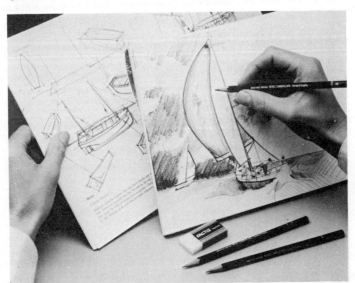

Figure 9.7b Graphite in action. Layers of graphite slide easily from the pencil onto paper.

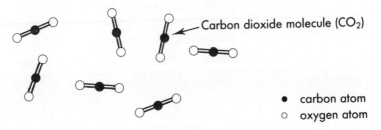

Carbon dioxide molecule (CO_2)

● carbon atom
○ oxygen atom

Explaining the Properties of Chemicals 147

Questions

1. Compounds formed between metals and non-metals are usually ionic. Compounds formed between non-metals only are usually covalent.
   (a) Classify the following compounds as ionic or covalent:
   (i) Sodium chloride   NaCl
   (ii) Potassium nitrate KNO_3
   (iii) Ethane C_2H_6
   (iv) Copper oxide   CuO
   (v) Sulphur dioxide SO_2
   (vi) Water H_2O
   (vii) Tetrachloromethane CCl_4
   (viii) Calcium carbonate $CaCO_3$
   (ix) Barium sulphate $BaSO_4$
   (x) Ammonia NH_3
   (b) Which of the compounds above would conduct electricity
   (i) When solid?
   (ii) When molten (liquid)?

2. Quartz is made of silicon dioxide, SiO_2. It is a giant structure held together by covalent bonds. Part of this giant structure is shown below, drawn in two dimensions:

   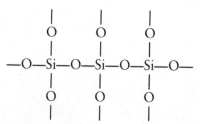

   (a) Explain carefully why quartz is a hard substance with a high melting point.
   (b) Explain whether or not you would expect molten quartz to conduct electricity.

3. Explain why
   (a) Calcium oxide has a high melting point.
   (b) Ammonia is a gas at room temperature.
   (c) Liquid sodium is chosen as a coolant in the fast-breeder nuclear reactor.
   (d) Argon is used to surround metals like titanium and aluminium when they are welded.
   (e) Silver is an expensive metal.
   (f) Aluminium alloys are used instead of pure aluminium to make aircraft bodies.

PART C

The Chemistry of Food Production

We may have to learn to live without fossil fuels and we may have to learn to live without many materials. We could obviously never live without food. We need food to provide us with energy and we need it to build up our bodies.

Chemistry can help at all stages of producing food. We can use chemicals to give us the right soil conditions for seeds to grow. We can use fertilizers to increase the size of our crops, and we can use chemicals to fight off crop diseases and pests. Chemistry is essential for producing our food today.

Lush grain crop. The size and quality of a crop can be improved by using chemicals.

10 Food Production

The World's Food Problem

There are over 4000 million people in the world to feed today. Each year this number grows by an extra 80 million, which is more than the whole population of the U.K. Over the past 30 years we have managed to increase the world's food supply each year, keeping up with the rising world population. This chapter is about the way in which chemists have helped to do this.

We produce just enough food today to feed the whole world, but still there are people starving. People in developed countries, especially North America and Europe, eat more food than they need. This leaves less food for the others, who cannot afford to buy enough. The World Health Organization (WHO) estimates that half the children in developing countries do not get enough food.

Growing food

About half the world's working population is involved in farming. In some countries, as much as 90% of the population is needed to grow food for everyone. In the U.K., only about 2%, or one person in fifty, takes part in farming. Even with this small number, in a small country, over half the food which is needed can be grown. The rest is bought from abroad. The main reason so much food can be grown in the U.K. is because extra energy is put into farming, apart from simple muscle power. Most of this energy comes from fossil fuels. Fossil fuels are used to run tractors and other machinery, and to make chemicals like fertilizers and weed-killers.

There are many ways of improving crops across the world. These include breeding better plants or animals, killing the pests which destroy crops, getting water to land which does not have it, and using photographs taken from satellites or aircraft to look at the way that land is used. We can also increase crops by adding fertilizers to the soil. We must add the right amounts of the right chemicals, so we need to know which chemicals are most needed by plants and animals.

The Chemicals in Food

Food is a mixture of many different chemicals. We use these chemicals to give us energy, and to build up our bodies. Altogether our food must contain the correct amounts of about 40 different

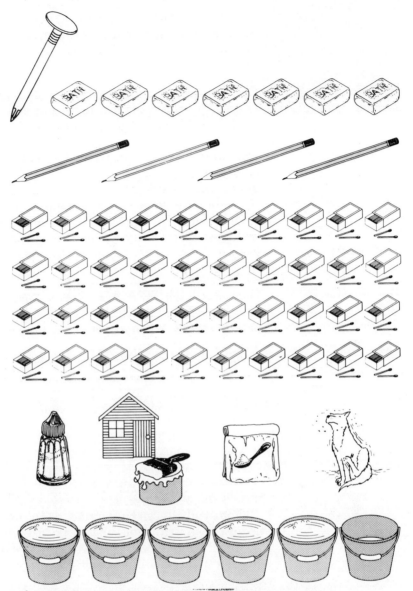

Figure 10.1 Rearranging the chemicals in an average human body.

chemical elements. To give you some idea, the chemicals in an average-sized body could be used to make one medium-sized nail, seven bars of soap, some graphite pencils, two thousand match heads, one medium-sized dose of salts, enough lime to whitewash a small shed, half a pound of sugar, enough sulphur to treat a dog for fleas, and about five and a half buckets of water (*figure 10.1*).

The most important foods are *carbohydrates* (starch and sugars), *fats* and *proteins*. The main use of carbohydrates and fats is to give us useful energy. We "burn" these foods by using oxygen, which we breathe in. This is known as **respiration**. We react the foods with oxygen, in a complicated way, and breathe out carbon dioxide and water. Proteins are needed to build up all the parts of our bodies. In addition to these main foods, we need small amounts of vitamins and many other chemicals to keep us alive.

Building foods from simpler chemicals

Carbohydrates and fats are compounds of carbon, hydrogen and oxygen. Proteins contain nitrogen as well.

The carbon and oxygen in foods come originally from carbon dioxide. There is plenty of that in the air. The hydrogen comes from water. If extra water is needed, it can often be brought to the land by irrigation (*figure 10.2*). Irrigated land now provides a quarter of the world's food. The nitrogen in proteins comes from the air or from simple nitrogen compounds in the soil.

Only plants can turn all these simple chemicals into useful ones which we can eat as foods. Humans and other animals could not live without the chemicals made by plants.

Photosynthesis

Plants use the energy from sunlight to build up simple chemicals into larger chemicals which we can use as foods. This is called **photosynthesis**. When we get energy from foods, we are really using stored solar energy.

Photosynthesis involves many different chemical reactions. Overall, carbon dioxide and water are converted to carbohydrates like glucose and starch. Chlorophyll, the green pigment in plants, is the catalyst which helps these reactions to take place.

A basic equation for photosynthesis can be written:

$$\text{Carbon dioxide} + \text{Water} \xrightarrow{[\text{chlorophyll}]} \text{Carbohydrate} + \text{Oxygen}$$
$$\text{(e.g. glucose)}$$
$$6\,CO_2 + 6H_2O \longrightarrow C_6H_{12}O_6 + 6O_2$$

Photosynthesis can be described as an endothermic process because energy from the sun is being taken in by the plant.

Notice that oxygen is produced during photosynthesis. We rely on plants to provide our oxygen to breathe, as well as our food to eat.

If we cut down trees faster than they can grow, there is a danger that we will upset the balance of oxygen in the atmosphere. Many people today are concerned that the tropical rain forest in places like South America is being rapidly destroyed.

The carbon cycle

All living organisms are based on compounds of carbon. This is why the chemistry of carbon compounds is called organic chemistry.

The movement of carbon from carbon dioxide in the air through living creatures and back to the air again is known as the carbon cycle. Photosynthesis is the vital link between carbon in carbon dioxide and carbon in plants and animals (*figure 10.3*).

Figure 10.2 The National Water Carrier in Israel. This water channel is 130 km long and carries water from the Sea of Galilee, Israel's only fresh water reservoir, to the deserts in the south of the country. The deserts are now being turned into land which can grow crops.

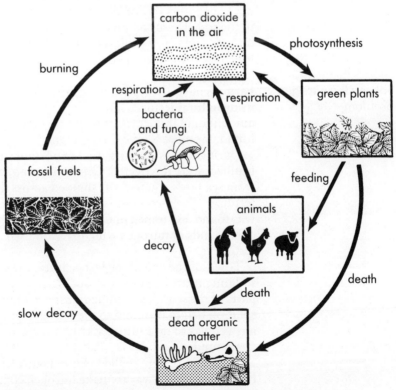

Figure 10.3 The carbon cycle. Carbon, contained in carbon dioxide, passes from the air into living organisms and eventually back to the air again.

Food Production 153

Extra chemicals for growing plants

Carbon dioxide and water are the basic chemicals which plants use in photosynthesis. The other chemicals which plants need are usually present in the soil. Plants take these chemicals from the soil when they grow. The chemicals which they use most are compounds of nitrogen, phosphorus and potassium. Nitrogen is especially important, because it is used for making proteins. When a crop is harvested, the nitrogen, phosphorus and potassium compounds go with it. Unless they are replaced in the soil, the next crop will be smaller.

For centuries, farmers have added these chemicals back to the soil, using fertilizers like manure. In the last hundred years, we have learnt how to make fertilizers from simple chemicals. Using these modern fertilizers, it is possible to double the value of many crops.

The Nitrogen Problem

There is plenty of nitrogen in the air, but few plants and no animals can use it directly. Most plants have to take in compounds of nitrogen through their roots.

A few types of plants, called legumes, can "fix" nitrogen directly from the air and make use of it. These plants, which include clover, peas and beans, have bacteria in their roots which do this for them. These bacteria are called nitrogen-fixing bacteria.

Other plants rely on compounds of nitrogen which are already in the soil. These compounds are used up quickly, unlike carbon dioxide and water. Unless they can be replaced we cannot grow so much food and more people starve.

Solving the nitrogen problem

Farmers have always put nitrogen compounds back into the soil by using natural fertilizers, including manure. There is only a limited amount of manure available. If other nitrogen fertilizers can be found, far more food can be grown.

During the last century, two important supplies of nitrogen fertilizer were discovered. One was guano, which is the droppings from sea birds. Guano was shipped across the seas to be used as a fertilizer. The second supply was sodium nitrate. This compound was found and mined in Chile and Bolivia. Nitrates are simple compounds of nitrogen which are still used in many modern fertilizers.

Towards the end of the last century, the supplies of guano and sodium nitrate began to run out. Scientists predicted that many parts of the world, including Europe, would face starvation. A solution had to be found to the nitrogen problem. The obvious answer was to take nitrogen from the air and turn it into useful compounds. Nitrogen is so unreactive that this is difficult to do.

A German scientist called Haber finally solved the problem in 1913. He found a way to make nitrogen combine with hydrogen. In this way he made a simple compound called ammonia. Ammonia can be used directly as a fertilizer, or it can be turned easily into other fertilizers.

Making Nitrogen Fertilizers

Making ammonia—the Haber Process

A modern factory making fertilizers still uses the method which Haber invented. It is known as the Haber Process. In the U.K. alone, over a million tons of nitrogen are turned into fertilizer each year using this process.

The nitrogen which is used comes straight from the air. It costs nothing, and there is plenty of it. The hydrogen comes from North Sea gas (methane) and steam, which are reacted together in a special way. Some other countries use oil instead of natural gas to supply their hydrogen.

The nitrogen and hydrogen are mixed together and made to react. It is so difficult to make nitrogen react that a high pressure (250 atmospheres) and quite a high temperature (450°C) are needed. Even then the reaction is slow, so a catalyst of iron is used to speed it up:

$$\text{Nitrogen} + \text{Hydrogen} \xrightleftharpoons{Fe} \text{Ammonia}$$
$$N_2 + 3H_2 \rightleftharpoons 2NH_3$$

Only a small amount of ammonia is made each time the nitrogen and hydrogen are sent past the catalyst. This is why the arrows in the equation show the reaction going both ways. The ammonia is taken out by condensing it to a liquid, and the nitrogen and hydrogen are taken round again. This is called *continuous flow*, because the gases are always flowing past the catalyst. Continuous flow is used for making many other chemicals apart from ammonia.

The reaction is exothermic, but this heat is not wasted. It is used to heat up the nitrogen and hydrogen before they reach the catalyst. Heat energy is too expensive to waste in a chemical factory.

A diagram of the Haber process is shown in *figure 10.4*.

Ammonia as a fertilizer

Ammonia can be used as a fertilizer, although little is used directly in the U.K. Ammonia is difficult to handle because it is a gas at normal temperatures and pressures. Sometimes the ammonia is liquefied under pressure and then injected into the soil.

Most of the ammonia which is produced is reacted with other chemicals to make solid fertilizers. These are easier to transport and to spread on the land. The most common nitrogen fertilizer today is ammonium nitrate, made by reacting ammonia with nitric acid. This means that large amounts of nitric acid have to be made, in addition to the ammonia. Modern fertilizer factories contain a nitric acid plant to do this as well as the ammonia plant.

The main uses of ammonia can be seen in *figure 10.6*.

Making nitric acid

Nitric acid is made from ammonia. Nitric acid HNO_3 contains oxygen, while ammonia NH_3 does not. Making nitric acid from ammonia therefore involves oxidation.

In the first part of the process, ammonia is reacted with oxygen. This is done by passing ammonia and air over a catalyst of platinum. Nitrogen monoxide gas is formed.

Food Production 155

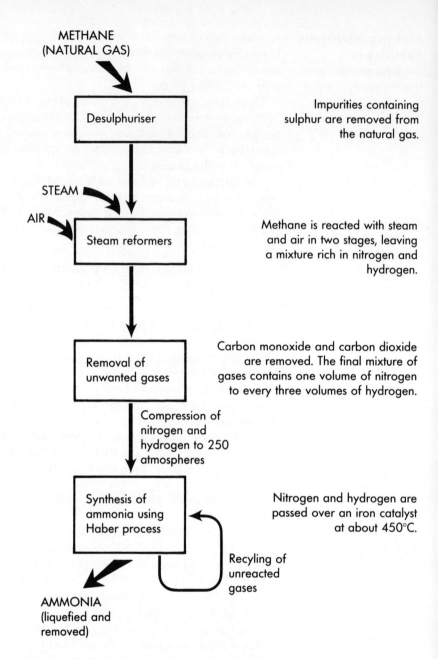

Figure 10.4 Making ammonia using the Haber process.

Ammonia + Oxygen → Nitrogen monoxide + Water
4NH₃ + 5O₂   4NO + 6H₂O

This reaction is exothermic. The heat which is given out is stored by making steam which can be used in other parts of the fertilizer factory.

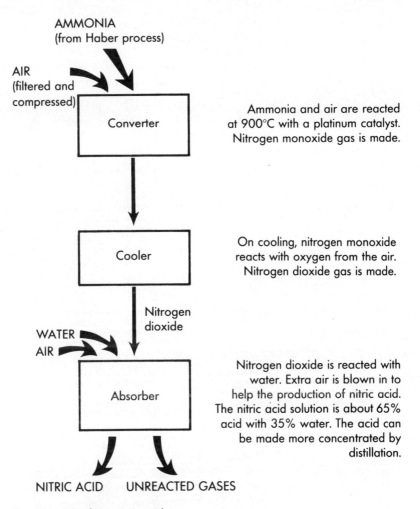

Figure 10.5 Making nitric acid.

When the nitrogen monoxide gas is cooled down, it reacts with more oxygen to make nitrogen dioxide gas:

Nitrogen monoxide + Oxygen → Nitrogen dioxide
$2NO + O_2 \qquad 2NO_2$

Nitric acid itself is made by reacting the nitrogen dioxide gas with water. Extra air is needed to make sure that as much nitrogen dioxide as possible is turned into nitric acid. If extra air is not used, a mixture of nitric acid with nitrous acid HNO_2 is made instead.

Nitrogen dioxide + Water + Oxygen → Nitric acid
$4NO_2 + 2H_2O + O_2 \qquad 4HNO_3$

A diagram of this process is shown in *figure 10.5*.

The last reaction is another example of the chemistry of non-metal oxides. Non-metal oxides, including nitrogen dioxide, form acids when they dissolve in water.

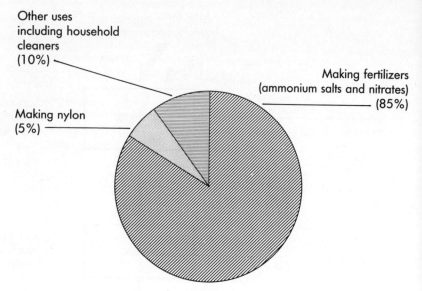

Figure 10.6 The uses of ammonia, NH$_3$.

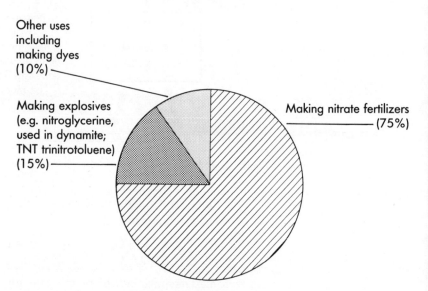

Figure 10.7 The uses of nitric acid, HNO$_3$.

A small amount of the nitrogen dioxide cannot be turned into nitric acid. It is expelled into the atmosphere through a tall chimney. It can cause pollution because it is acidic (p.23). In older nitric acid plants, a brown plume of nitrogen dioxide can be seen above the chimney. Newer factories have better controls, and the escaping gas is no longer brown and visible.

Nitric acid has many uses outside the fertilizer industry. The main ones are shown in *figure 10.7*.

Figure 10.8 Part of the Billingham site, headquarters of ICI's Agricultural Division. The ammonia plants and nitric acid plants are close together on the site.

Billingham was chosen as a site for making fertilizers because it has good links by road, rail and sea, and because it is close to water (the Teesdale reservoirs), coal, a new power station, and workers from nearby Stockton and Middlesbrough.

Some plants on the site are used to make important byproducts. The methanol, urea and liquid carbon dioxide plants shown in the photograph are only part of the complex. These three plants use carbon dioxide which comes from the ammonia plant.

1 Ammonia plants   2 Nitric acid plants   3 Liquid carbon dioxide plants
4 Urea plant   5 Methanol plants

Nitric acid plants and ammonia plants are often built next to each other. You can get an idea of the large scale of this chemical industry by studying *figure 10.8*.

Food Production 159

Ammonium nitrate— the final product

Ammonia and nitric acid are reacted together to make the final nitrogen fertilizer—ammonium nitrate:

Ammonia + Nitric acid → Ammonium nitrate
NH_3 + HNO_3 NH_4NO_3

The heat given out in this reaction is enough to melt the ammonium nitrate. The liquid ammonium nitrate is pumped to the top of a tower about 100 m high, and then sprayed down it. As it falls down the tower it breaks into drops. The drops cool down and form solid granules. These solid granules of fertilizer can be packed into bags and sold to farmers (*figure 10.9*).

Figure 10.9 The fertilizer manufacturing section of the Billingham site. Chemicals made in other parts of the site are combined here to make the final fertilizers.

1 "Compound" fertilizer plant (NPK fertilizers)
2 Nitrogen fertilizer plant with two "prilling" towers
3 Packaging and storage area for fertilizers

The nitrogen cycle

The Haber process is an important link between nitrogen in the air and nitrogen compounds, including proteins, in living organisms. Without modern fertilizers, based on the Haber process, we would grow far less food in the world.

The movement of nitrogen from the air through living creatures and back to the air again is known as the nitrogen cycle (*figure 10.10*).

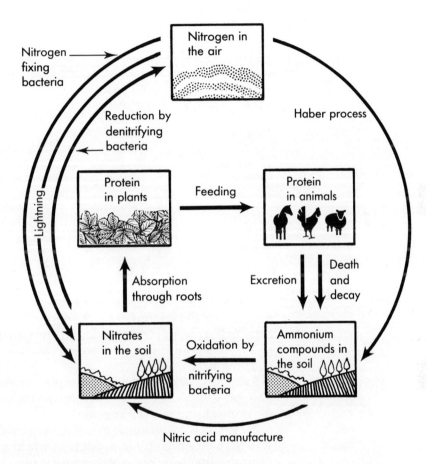

Figure 10.10 The nitrogen cycle. Nitrogen from the air passes through living organisms and eventually back to the air.

Acids, Bases and Salts

The reaction of nitric acid with ammonia to form ammonium nitrate is an example of an important and common sort of reaction. When **acids** react, compounds called **salts** are formed. Ammonium nitrate is an example of a salt, because it can be made from nitric acid. The chemicals which react with acids to make salts are called **bases**. Ammonia is a base, because it will react with acids to form ammonium compounds, which are salts.

You should think of a base as the opposite of an acid. Whenever an acid and a base react together they **neutralize** each other, and a salt is formed.

Acid + Base → Salt

Different acids and their salts

Each acid produces its own family of salts when it reacts with different bases. The salts which are made from some common acids are shown in *table 10.1*.

Acid	Family of salts	Example of salt
Hydrochloric acid	Chlorides	Sodium chloride ("salt")
Sulphuric acid	Sulphates	Magnesium sulphate (Epsom salts)
Nitric acid	Nitrates	Ammonium nitrate (fertilizer)
Phosphoric acid	Phosphates	Calcium phosphate (bone)
Carbonic acid	Hydrogencarbonates	Sodium hydrogencarbonate (baking soda)
Carbonic acid	Carbonates	Calcium carbonate (chalk)
Stearic acid	Stearates	Sodium stearate (soap)

Table 10.1 Some acids and their salts

Bases

Apart from ammonia, the common bases are metal oxides and metal hydroxides. By contrast, the non-metal oxides, including carbon dioxide, sulphur dioxide and nitrogen dioxide, are acidic (p.23). This is an important chemical difference between metals and non-metals.

Bases can always be used to neutralize unwanted acids. One example is the use of Milk of Magnesia to neutralize excess acid in the stomach and relieve indigestion.

Magnesia is a suspension of magnesium oxide, a metal oxide, in water. The stomach contains dilute hydrochloric acid. A neutralization takes place between these two chemicals.

Magnesium oxide + Hydrochloric acid → Magnesium chloride + Water
$MgO + 2HCl$ $MgCl_2 + H_2O$
base + acid   salt + water

Alkalis

Any base which dissolves in water is called an **alkali**. This means that all alkalis are bases, and will react with acids to produce salts. The cheapest and most commonly used alkali is sodium hydroxide, which is described on page 134. Sodium hydroxide will neutralize hydrochloric acid to form a salt called sodium chloride.

Sodium hydroxide + Hydrochloric acid → Sodium chloride + Water
$NaOH + HCl$ $NaCl + H_2O$

Sodium chloride is used so commonly in our homes and factories that we just call it "salt".

There are only a few common alkalis, apart from sodium hydroxide solution. Ammonia solution and calcium hydroxide solution (limewater) are two of them.

Making Phosphorus Fertilizers

Many modern fertilizers are "compound" fertilizers, containing nitrogen, phosphorus and potassium. The most common chemical containing phosphorus which is found naturally is calcium phosphate rock. Unfortunately it is insoluble in water so it does not dissolve in the soil, and plants cannot use it. It has to be turned into a soluble chemical which plants can use. This is usually ammonium phosphate. It is made by the same sort of reaction used to make ammonium nitrate (p.160), except that phosphoric acid is reacted with the ammonia instead of nitric acid.

Mining calcium phosphate

Most of the calcium phosphate used in the U.K. is imported. A small amount, about 5%, comes from our steel industry. This is possible because calcium phosphate is made when phosphorus impurities are removed from pig-iron to make steel (p.89). Three of the main suppliers of calcium phosphate to the U.K. are the African countries of Morocco, Senegal and Tunisia. The trade is very important to them, because calcium phosphate rock is one of their main exports.

The rock is crushed near the mine and then shipped to the U.K. Almost all the phosphate rock is used for making fertilizers, although some is turned into detergents (p.202).

In order to make ammonium phosphate fertilizer, phosphoric acid has to be made from the phosphate rock. This is done by reacting the crushed rock with sulphuric acid.

Calcium phosphate + Sulphuric acid →
                    Calcium sulphate + Phosphoric acid

In the U.K., more sulphuric acid is used for this purpose than for any other. A large fertilizer factory will have a sulphuric acid plant, in addition to its ammonia and nitric acid plants.

Making sulphuric acid—the Contact Process

Most sulphuric acid in the U.K. is made from the element sulphur itself. About 20 kg of sulphur are used for every man, woman and child in the country each year. Even so, the total is less than the amount which falls on the land as acid rain. Almost all the sulphur used in the U.K. is turned into sulphuric acid.

The sulphur comes from two main sources. One is from underground deposits of sulphur. There are sulphur mines in some parts of the U.S.A. Sulphur also comes from fossil fuels. Fossil fuels always contain some sulphur (p.16). Most of it is removed before the fuel is burnt, because otherwise there would be more pollution

Food Production

from sulphur dioxide gas. North Sea oil and gas do not contain much sulphur, so sulphur for the U.K. is bought from abroad. About half the sulphur comes from sulphur mines and the other half from fossil fuels.

Sulphur is converted into sulphuric acid in four steps:

1. Sulphur is oxidized by burning it in air to form sulphur dioxide:

Sulphur + Oxygen → Sulphur dioxide
$S + O_2$ SO_2

2. A mixture of sulphur dioxide and more air is cleaned and dried. It is then pumped into a reaction chamber containing a catalyst of vanadium (V) oxide, at a temperature of about 450°C and atmospheric pressure. In this reaction, sulphur dioxide is oxidized to form sulphur trioxide:

Sulphur dioxide + Oxygen → Sulphur trioxide
$2SO_2 + O_2$ $2SO_3$

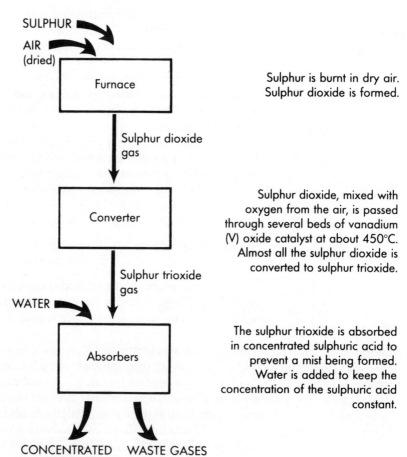

Figure 10.11 Making sulphuric acid using the Contact process.

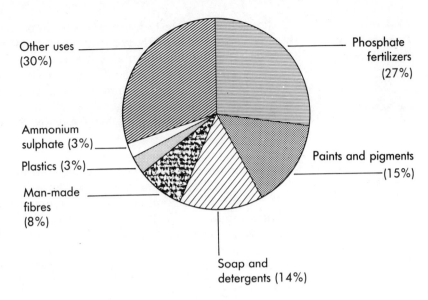

Figure 10.12 The uses of sulphuric acid, H_2SO_4, in the U.K. (1980).

3. Sulphur trioxide is an acidic oxide, because it is the oxide of sulphur, a non-metal. When it reacts with water, it forms sulphuric acid. This cannot be done immediately in a factory, because the sulphuric acid is formed as a fine mist which is poisonous and difficult to handle. Instead, the sulphur trioxide is dissolved in concentrated sulphuric acid itself. The solution formed is known as oleum.

4. Water is added to the oleum. The water reacts with the dissolved sulphur trioxide to form concentrated sulphuric acid:

Sulphur trioxide + Water → Concentrated sulphuric acid
$$SO_3 + H_2O \quad\quad H_2SO_4$$

A diagram of this process is shown in *figure 10.11*.

Although the major use of sulphuric acid is for making phosphate fertilizers, it is widely used for other purposes (*figure 10.12*).

Ammonium phosphate—the phosphorus fertilizer

Ammonium phosphate is made in another neutralization reaction, just like ammonium nitrate. This time the ammonia is neutralized by phosphoric acid:

Ammonia + Phosphoric acid → Ammonium phosphate
(base)   (acid)   (salt)

In many cases the nitric acid and phosphoric acid are neutralized together by the ammonia. This produces a mixture of ammonium nitrate and ammonium phosphate which can be made into granules and packed into bags.

Making Potassium Fertilizers

Potassium is the last of the three most important elements which plants need for growth. Potassium fertilizers are much easier to make than fertilizers containing nitrogen or phosphorus. The reason is that potassium chloride, which dissolves easily in the soil, can be found and mined underground. It only needs to be dug up, purified, and put into bags.

Potassium chloride is a white solid, just like sodium chloride. The potassium chloride used in the U.K. comes from near Whitby in Yorkshire. The mine contains a mixture of potassium chloride and sodium chloride. The potassium chloride is separated from the ordinary salt, dried and then packaged.

A typical compound fertilizer, containing ammonium nitrate, ammonium phosphate and potassium chloride is known as an NPK fertilizer (*figure 10.13*). Work out why it has this name.

Many different chemicals are involved in making a compound fertilizer. The overall process is shown in *figure 10.14*.

Energy and Fertilizers

Food gives us energy, but people have to use energy in order to grow food. In the U.K., only half the energy which goes into growing food is returned as food energy. This seems wasteful. An African bushman can get eight times more energy out of the food than he puts in. The difference is that a British worker grows over a

Figure 10.13 A typical bag of "compound" or "NPK" fertilizer.

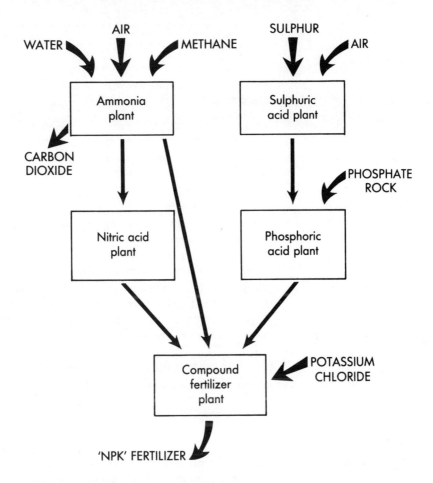

Figure 10.14 Making a compound fertilizer.

thousand times as much food as the bushman on each acre of land and can therefore grow food for far more people. The reason this can be done is that so much extra energy is put in apart from muscle power. Fossil fuels are used to drive the tractors and other machinery. They are also used for making fertilizers and pesticides. Even so, the farming industry in the U.K. only uses about 4% of the total supply of energy. It is also quite a good way of using our precious fossil fuels. For example, a bag of ammonium nitrate fertilizer (50 kg) contains the same amount of energy as 30 litres of petrol. The bag of fertilizer will grow enough grain for 530 people for one day. The petrol will take just one car a distance of about 200 miles. So, which do you think is the better way of using fossil fuels?

The prices of different fuels, fertilizers and foods which have the same energy value are compared in *table 10.2*.

Energy-giving chemical	Approximate cost (1983)
5 litres petrol	£2.00
7 kg coal	£1.00
8 kg ammonium nitrate fertilizer	£1.20
15 kg bread	£7.70
60 litres milk	£22.40
19 kg steak	£125.00

Table 10.2 These chemicals all provide the same amount of energy when used. Which do we value more, fuel or food? Why?

Even though farming is a good use of fossil fuels, there are people who suggest that they could be used more carefully. For example, it takes much more fertilizer and land to grow cattle and sheep for meat than to grow the same amount of food energy and protein as grain and vegetables. Some people are vegetarians because they think it is wasteful to produce meat.

Many countries cannot afford to spend as much as Western countries on fertilizers and fuel for farm machinery. They desperately need cheaper supplies of fertilizers and energy. One possibility is to build small biogas plants, described in Chapter 3 (p.50). Using animal dung, human waste and the remains of crops, a rich fertilizer can be made.

Pollution from Fertilizers

In the days before artificial fertilizers were made, farmers only put natural chemicals on the land. They used animal dung and vegetable remains, as well as substances like bone meal for phosphorus (bone contains calcium phosphate). All these fertilizers are still used today. The extra chemicals are simply used to grow bigger crops to feed the enormous world population.

Over 6 million tons of man-made fertilizer are spread on the land in the U.K. each year. Large amounts of this wash into rivers, lakes and water supplies. The worst problems are caused by nitrates and phosphates.

Nitrates and phosphates are used by water plants, including algae, as food. If there is too much nitrate or phosphate, the algae grow rapidly and cover the water. Underwater plants can no longer obtain sunlight, so they die. They are decomposed by bacteria, which use up the oxygen in the water. Many other creatures, including fish, start to die because they cannot get enough oxygen. This process is called eutrophication. It makes the water lifeless and smelly. The Norfolk Broads in the U.K. are affected in this way. Lake Erie in Canada is a more serious example. Even if the pollution stopped, it is thought that the lake would take a century to recover.

Nitrates are dangerous to humans, especially babies, in water supplies. They can stop the blood from carrying oxygen properly, as carbon monoxide can. In the drought of 1976 in the U.K., the concentration of nitrates in the water was much higher than usual, because there was so little water. The government warned that babies should not be fed with ordinary drinking water.

We could reduce this pollution if we used less man-made fertilizer. Some people think that we should only use natural fertilizers, called "organic" fertilizers, but if every farmer did this we would produce much less food.

Acids and Alkalis in Soils

Crops will not grow properly if the soil is too acidic or too alkaline, even if they have plenty of fertilizer, water and sunshine. In the U.K., too much acid is the main problem. If this acid was neutralized properly, perhaps 20% more food could be grown.

The pH scale

Acidity, which means the concentration of acid, is measured on a scale called the **pH scale**.

pH7 is neutral—neither acidic nor alkaline. Ordinary water has a pH of 7.

Acidic solutions have pH numbers below 7. *The lower the number, the more concentrated the acid.* The pH of the hydrochloric acid in your stomach is about 2. The pH of hydrochloric acid on the laboratory bench is about 1.

Alkaline solutions have pH numbers greater than 7. The higher the number, the more concentrated the alkali. Laboratory sodium hydroxide has a pH of 13.

Indicators

Using an indicator, it is possible to tell if a solution is acidic or alkaline. **Indicators** are chemicals which turn different colours in acids or alkalis.

A commonly used indicator is litmus, which turns red in acids and purple in alkalis. In soil testing, a universal indicator is used. This is a mixture of many different indicators, which shows the exact pH of the soil. The colours of universal indicator in solutions of different pH are shown in *figure 10.15*.

Universal indicator colour	RED			O	Y	G	B	I		VIOLET			
pH	1	2	3	4	5	6	7	8	9	10	11	12	13	14
Solution	Strongly acidic			Weakly acidic			N	Weakly alkaline			Strongly alkaline			

Figure 10.15 The colour of universal indicator in different solutions.

Solution	pH
Dilute hydrochloric acid	1
Dilute nitric acid	1
Dilute sulphuric acid	1
Citric acid (oranges and lemons)	2
Ethanoic acid (vinegar)	3
Tartaric acid (grapes)	3
Carbonic acid	6
Fresh cow's milk	6.5
PURE WATER	7
Sodium chloride (salt)	7
Sucrose (sugar)	7
Ethanol (alcohol)	7
Sodium hydrogencarbonate (baking soda)	8.5
Ammonia	10
Sodium carbonate (washing soda)	11.5
Calcium hydroxide (limewater)	12.5
Sodium hydroxide	13

Figure 10.16 The pH values of some common solutions.

pH and crops

Most crops grow best in a soil which is slightly acidic (pH 6 to 6.5). If the pH becomes less than 5, then there are usually problems.

Different crops need different pH values in the soil for their best growth (*table 10.3*).

Crop	pH range for best growth
Sugar beet	7.0 to 7.5
Wheat	6.0 to 7.5
Turnips	5.5 to 7.0
Oats	4.8 to 6.3
Swedes	4.7 to 5.6

Table 10.3 Ideal pH values for different crops

Some soils are naturally very acidic. The soils on the Yorkshire Moors have a pH of about 3. It is possible to grow forests on these soils but not ordinary crops.

If man-made fertilizers are used on ordinary soils, the soil usually becomes more acidic. The extra acid has to be neutralized if the best crops are to be grown.

Lime—to neutralize the extra acid

The unwanted acid is neutralized by adding a base to the soil. A suitable cheap base is lime (calcium oxide), which can be made from limestone (calcium carbonate).

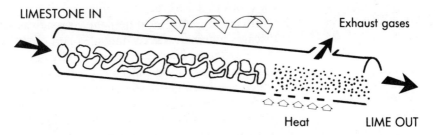

Figure 10.17 A rotating lime kiln. The kiln is a large cylinder which slopes gently and rotates at one revolution per minute. Limestone is loaded in at one end. It dries out as it moves down to the heated end, where it is chemically changed into lime.

Lime is made on a large scale in lime-kilns (*figure 10.17*). The limestone is heated in a special furnace. It breaks down giving lime and carbon dioxide gas:

Calcium carbonate → Calcium oxide + Carbon dioxide
$CaCO_3$ $CaO + CO_2$

Sometimes the calcium oxide is used directly. It is called quicklime. In other cases, the quicklime is reacted with water to form slaked lime, which is calcium hydroxide. This reaction produces a large amount of heat, enough to turn some of the water into steam.

Calcium oxide + Water → Calcium hydroxide
(lime) (Slaked lime)
$CaO + H_2O$ $Ca(OH)_2$

Calcium carbonate itself, in the form of chalk, can also be used to neutralize acids in the soil, since all carbonates react with acids and neutralize them. It is often better to use chalk instead of lime, because it does not dissolve in water and does not wash off the land so quickly.

Quicklime and slaked lime are both alkalis because they are bases which dissolve in water. Limewater, a laboratory alkali, is just slaked lime which has been diluted.

Lime is added not only to neutralize the extra acid. It also helps the drainage of water through the soil. This is especially important in thick clays, which would become very sticky and waterlogged if they did not have lime added from time to time.

Some soils in the U.K. have pH numbers greater than 7, which means that they are alkaline. This is most likely to happen in parts of the country where there is chalk or limestone beneath the soil. These soils have to be made more acidic if crops are to grow well.

Pesticides

Pesticides are chemicals which kill pests.

Pest control is vital for food production. In the world as a whole, a third of all crops are destroyed by pests and diseases. In Asia, Africa and South America the amount is even higher. Over 40% of the crops in these areas are ruined.

Insects cause the most damage, as much as all other pests and diseases together. Insects damage plant crops by biting and chewing them. They can also carry virus diseases to plants and cause infections in animals like cows and sheep. The chemicals which are used against insects are called insecticides.

Weeds can also damage crops. They may stop the crop from growing well and spoil the harvest. Some weeds are poisonous, and can make a crop useless as food. The chemicals which are used against weeds are called herbicides.

There are many other types of pests, apart from insects and weeds. The most important of these are the fungi, including mildew. Fungicides are used to deal with these pests.

Insecticides

Half the pesticides sold in the world are **insecticides**. DDT was the first man-made insecticide. It was made in 1874, although it was little used until after the Second World War. DDT stands for dichloro-diphenyl-trichloroethane.

It is a complicated chemical (*figure 10.18*), like most pesticides. DDT is an organic chemical and it contains chlorine. It belongs to a family of insecticides called *organochlorines*. BHC (benzene-hexachloride), aldrin and dieldrin are also organochlorines.

DDT has been of great use to us. It has saved millions of lives by killing the mosquitoes which carry malaria. Unfortunately, chemicals like DDT have harmful effects as well.

DDT is a long-lasting chemical, it finds its way into animals through the food chain (*figure 10.19*). The amount of DDT gradually builds up, especially in the fatty parts of the body, until the animal is poisoned. Sparrow hawks and peregrine falcons are just two types of bird which have been affected by this type of pesticide in the U.K. Many countries now ban or severely control the use of DDT and other organochlorine pesticides. Perhaps they should be banned completely.

A second group of insecticides consists of organic compounds which contain phosphorus. These are the *organophosphorus* insecticides. An example is malathion (*figure 10.18*). These are replacing the organochlorines like DDT because they are not as long-lasting. Even so, there are many problems with these insecticides. They can damage animal life. Insects also become resistant to them. Scientists have to find new chemicals to use or new ways of using them. If larger amounts of the old insecticide are used to overcome resistance, there is a risk of more serious pollution.

Insecticides are usually applied to crops by spraying. Back-packs, tractors or aircraft can be used to do this (*figure 10.20*).

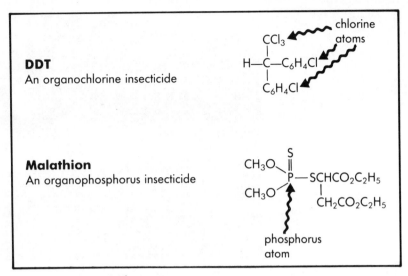

Figure 10.18 Insecticides.

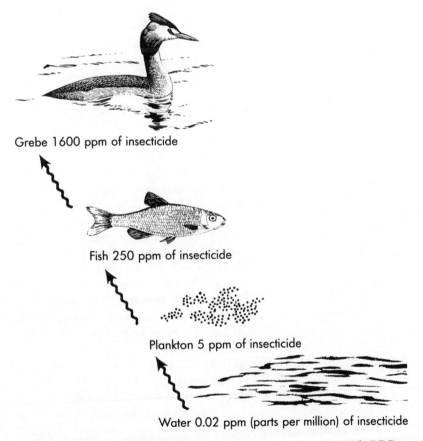

Figure 10.19 Clear Lake in California was sprayed with the insecticide DDD, similar to DDT. Many of the grebes living on fish in the lake were later found dead. They were poisoned by the insecticide which passed through the food chain from the water to plankton, fish and finally to the grebes.

Figure 10.20 Aerial spraying of the pesticide "Ambush" onto a cotton crop.

Herbicides

Herbicides can be divided into two groups. Some herbicides will kill all types of plant. They are known as non-selective herbicides. Other chemicals, called selective herbicides, will only attack certain plants.

Paraquat is an example of a non-selective herbicide (*figure 10.21*). It kills all green plants. It is especially useful because it becomes inactive after only a few hours in the soil. A whole field can be treated and then replanted. Paraquat also has its dangers. It is fatal if it is swallowed, and there is no known antidote. Paraquat, like other dangerous chemicals, should always be kept in a labelled container, well away from small children.

2,4,5-T is an example of a selective herbicide (*figure 10.21*). It is especially useful against tough weeds like nettles and brambles, and in protecting cereal crops. Few pesticides are argued about as much as 2,4,5-T. It contains traces of dioxin, which is a very poisonous chemical. Small doses of it are known to cause cancer and birth defects in animals. Dioxin is also found in Agent Orange, a chemical weapon used against the Vietnamese by the U.S.A. 2,4,5-T is now banned in many countries.

Fungicides

Fungi can cause great damage to crops. The famine in Ireland in 1845 was caused by a fungus called potato blight. A million people died of starvation when the potato crop failed.

Many **fungicides** contain copper compounds. An example is Bordeaux mixture, which contains copper sulphate and lime. This mixture was discovered by accident. A French wine grower put it on his vines to stop children damaging them. He then found that the chemicals stopped mildew and other diseases from attacking the vines as well.

Paraquat
A non-selective herbicide

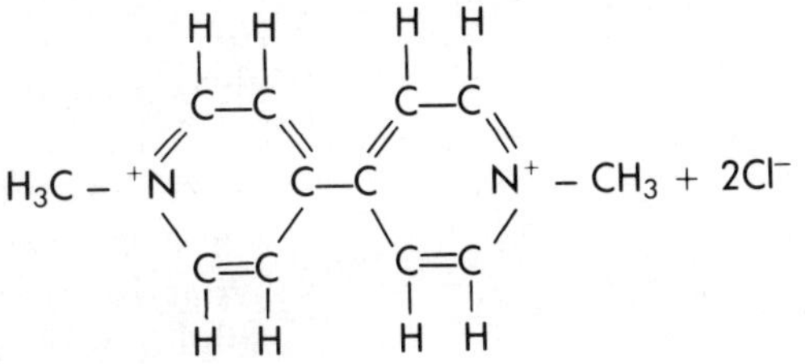

$$H_3C - {}^+N\bigcirc\!\!=\!\!\bigcirc N^+ - CH_3 + 2Cl^-$$

2,4,5-T
A selective herbicide which attacks broad-leaved plants

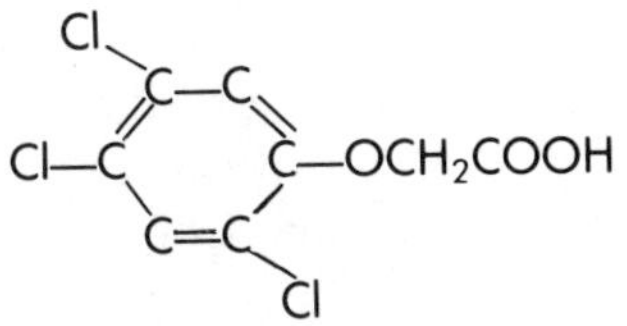

(structure: 2,4,5-trichlorophenoxyacetic acid — a benzene ring with Cl substituents and $-OCH_2COOH$)

Figure 10.21 Herbicides.

The future

The dangers of pesticides are now clear. Some pesticides will always be needed for obtaining good crops. The use of pesticides should be combined both with breeding crops which resist disease and with developing other biological methods of controlling pests. Proper crop rotation is also important.

Pesticides should be used in the smallest amounts possible. Pesticides which stay dangerous for years, like DDT, should not be used at all.

Scientists test thousands of new chemicals each year to see if they will work as pesticides. An ideal pesticide should only attack the pest itself, leaving other living things alone. It is very difficult to make such a pesticide. Only a few insects are pests and yet insecticides kill many types of insects and affect animals as well. There is much work left for chemists to do.

Questions

1. Photosynthesis can be described as "endothermic". Explain why.
2. The main gases in the atmosphere are shown in the table below:

Gas	Formula	Approximate percentage by volume of the atmosphere
Nitrogen	N_2	78%
Oxygen	O_2	21%
Noble gases (mainly argon)		1%
Carbon dioxide	CO_2	0.03%
Water vapour	H_2O	Small and variable

(a) Carbon dioxide is used by plants, but the amount of carbon dioxide in the air hardly changes. Explain why.

Food Production

(b) Explain why the amount of oxygen in the atmosphere is also constant.
(c) Why do fuels burn more slowly in air than in oxygen?
(d) What other substances would you expect to find in the atmosphere and why?

3 What do respiration, burning and rusting have in common?

4 Sulphur is the raw material for making sulphuric acid, used in the manufacture of phosphate fertilizers.

Much of this sulphur is obtained from underground deposits by a method known as the Frasch process.

Read the passage which follows and then answer the questions about it.

The sulphur is usually about 150 metres below the surface, underneath soft rocks like clay. Poisonous gases like sulphur dioxide and hydrogen sulphide are found with the sulphur. A Frasch pump is used to bring the sulphur to the surface. It consists of three pipes inside each other.

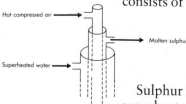

Sulphur melts at 115°C. The outer pipe contains superheated water at 170°C, which is pumped down to melt the sulphur. Superheated water is water heated under pressure so that its temperature is above 100°C. Molten sulphur comes up the middle pipe. Hot compressed air is pumped down the small centre pipe to help to push the molten sulphur to the surface.

(a) Why is the water heated under pressure?
(b) Why are ordinary mining methods not used?
(c) Why is the sulphur forced up the middle pipe?
(d) Why is the compressed air hot?
(e) Why do you think that the sulphur made by this method is very pure?

5 State whether each of the following chemicals is an acid, a base or a salt.
(a) H_2SO_4 (f) HNO_3
(b) KNO_3 (g) Fe_2O_3
(c) NH_4NO_3 (h) $CaSO_4$
(d) CaO (i) $NaOH$
(e) $CaCO_3$ (j) KCl

6 Write a short paragraph explaining the advantages and disadvantages of modern fertilizers.

7 After an insecticide has been used for a while, insects often become resistant to it. The ordinary dose of the insecticide no longer works as well. What can farmers and chemists do about this? What difficulties do you think they will meet?

PART D

Chemistry at Home

It is impossible to imagine life today without chemistry. Hundreds of useful chemicals find their way into our homes. Many of these, such as natural gas, metals, plastics and food, have been described in earlier chapters. This final section is about some of the more common chemicals which we use at home. Look around your house, especially the kitchen and bathroom, to find some of these chemicals. You should find soaps, detergents and bleaches for washing and cleaning. You can look out for drugs and other medicines, which are all chemicals. You should also be able to find chemicals used in cooking and, perhaps, in brewing and wine-making.

This section starts with a chemical which we take for granted—water.

Is the kitchen in the photograph a typical kitchen? If not, describe a typical kitchen and compare it with this one in terms of materials and other chemicals used.

11 Water Supplies

Water

Water is our most important chemical, after oxygen. People can stay alive for weeks without food, but they live only for about six days without water. Two-thirds of you is water. You need to take in about 2 litres of water every day from your food and drink. This replaces the water which you lose in sweat, in urine and by breathing.

Water in the world

Almost all the water in the world is salt water, found in the seas and oceans. This is not much use to us. We need fresh water for almost everything: for drinking, cooking, washing, and for our industries. Only 3% of the world's water is fresh. Most of this is frozen in the Arctic and Antarctic. Even so, there is enough fresh water in most countries at the moment.

How water is used

Water is taken for granted in the U.K. Each person uses about 150 litres of water at home every day. Every time you flush a lavatory, 9 litres of water goes down the drain (*table 11.1*).
    Water is not just used at home. Far more is used in industries. Some examples of these uses are shown in *table 11.2*.
    When all these uses are added together, each person in an industrialized country uses about 500 litres of water each day.

The Water Cycle

Water rarely stays in one place for long. It does this only if it is frozen in permanent ice or if it is deep under the earth. Some water in deep artesian wells has been there fore more than 20 000 years. Most water only stays in the same place for a few days. The average water molecule in the atmosphere stays there for 10 days.
    The movement of water on our planet is powered by the sun. Solar energy evaporates water into the air from rivers, lakes and oceans. Plants, which get energy from the sun, lose water into the air by a process called *transpiration*. The water collects into clouds. When the clouds rise, perhaps over hills and mountains, they cool down. Water drops are formed by condensation. The water then falls as rain, snow or hail. Some of the water falls straight back into the sea, but most of it falls on the land. It is then carried away by

Use of water	Amount used by each person each day
Washing and bathing	50 litres
Flushing the lavatory	50 litres
Laundry	15 litres
Washing up	15 litres
Watering garden and car washing	10 litres
Drinking and cooking	5 litres
	145 litres

Table 11.1   Using water at home

Use of water	Amount used
Making a family car	450 000 litres
Making a tonne of steel	200 000 litres
Making paper for one newspaper	200 litres
Making a bag of cement	180 litres
Making one litre of beer	10 litres

Table 11.2   Using water in industry

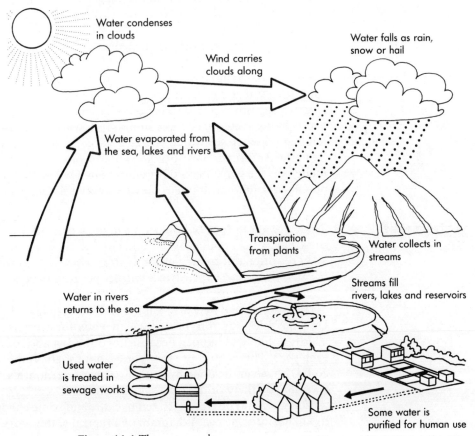

Figure 11.1   The water cycle.

Water Supplies   179

streams and rivers. Much of it eventually finds its way back to the sea, and the cycle starts again. This continuous movement of water is called the water cycle (*figure 11.1*).

Collecting water

Water supplies in the U.K. are taken mostly from rivers, lakes and reservoirs. There is normally plenty of water because so much rain falls on the U.K. Rain which falls on the mountain areas can be stored and piped to the big cities. In one year, 1976, people in the U.K. saw what happens in a very small drought. It stopped raining for two months. People had to stop watering grass and gardens, cleaning cars or working fountains. Even so, there was enough water for everyone. It rained again before many people had supplies to their houses cut off.

Many countries have to live with drought all the time, or with a sudden drought which lasts for months. Crops and livestock die, and searching for water becomes a major problem (*figure 11.2*). Countries in Africa and South America, together with Australia, have been badly affected by droughts in the 1980s. Droughts affect people as badly as they affect food crops. The drought in Ethiopia in 1972 killed at least 200 000 people.

In some countries fresh water is made by distilling sea water. This is called *desalination*, which means removing salt. It is expensive because so much heat is needed to boil off the water. The biggest desalination works in the world is in Hong Kong. It has never been used because the oil needed for it is too expensive. Only countries like Saudi Arabia and Kuwait, which have their own supplies of oil and gas, can afford to make water like this.

One suggestion is that icebergs could be towed to hot countries for drinking water. Icebergs contain fresh water, because the salt is left behind when sea water freezes. A large iceberg could be towed from the Antarctic to Australia in about a month. Even if half of it melted on the way, it could provide enough water for 4 million people for one year. It might only cost about £1 per person.

Purifying water

There are about four billion people in the world. One billion of these have to drink dirty water and two billion have no lavatories. Water is often not treated properly in less-developed countries because this is too expensive. Some people even have to wash in open sewers (*figure 11.3*).

Diseases carried by water kill five million people each year. The United Nations aims to provide clean water for everyone by 1990. This is unlikely to happen because it would cost about £25 billion each year. This is far more than the amount spent at the moment. It could, however, be done by using just over one month's military spending by the world.

Water should be treated to remove dirt and some chemicals, and to kill harmful germs. A diagram of a typical water works is shown in *figure 11.4*. The water is first filtered to remove large objects like

Figure 11.2 Searching for water in a dried-up river bed in Kenya. In the hilly areas of Kenya, women spend 90% of their time collecting water.

Figure 11.3 Washing in an open sewer in Calcutta, India. 60% of people in the Third World cannot easily obtain clean water.

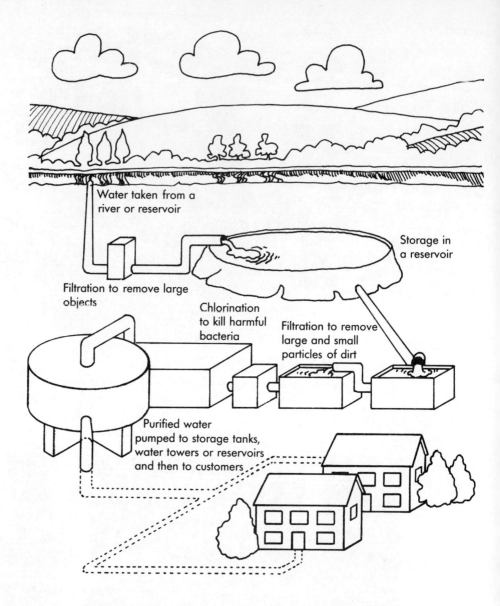

Figure 11.4 Water purification at a water works.

182 Chemistry in Use

driftwood. It is then pumped to a small reservoir or storage tower. When it is needed, the water is filtered through sand beds. Coarse sand is used first, to remove large pieces of dirt. Finer sand is used after this, to take out smaller particles of dirt. The water may have chemicals added if it is too "hard" or too "soft" (p.186). Some dangerous bacteria still remain after this treatment. Chlorine is added to the water to kill them. Chlorination of water supplies is especially useful to the poorer countries because it is a fairly cheap and easy way of purifying water.

After purification, the treated water is pumped to houses, factories and other buildings.

Treating sewage

Used water is often recycled because this saves water. It cannot be reused immediately, or put back into rivers, because it contains chemicals which can cause pollution. Used water contains human waste, which can contain harmful bacteria. The water also contains dirt and detergents from sinks, baths and washing machines, together with industrial waste.

The waste water is treated in a sewage works. A diagram of a sewage works is shown in *figure 11.5*. The raw sewage is first filtered to remove pieces of cloth and other large objects. It then flows slowly along special channels, so that any grit or sand can

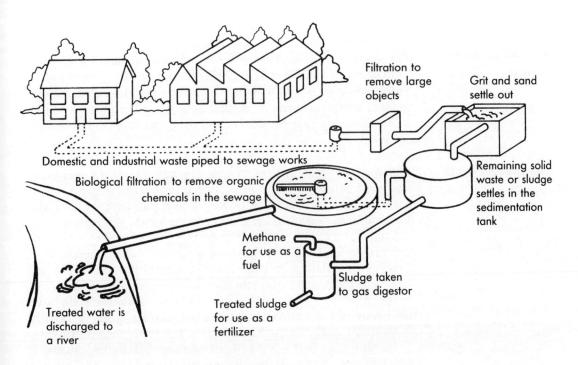

Figure 11.5 Treatment of waste in a sewage works.

settle out. Any remaining solid waste or sludge is allowed to settle in sedimentation tanks. The sludge is then pumped to a gas digester. Tiny organisms in the digester break down the sludge. They produce gases, including methane, which can be used as a fuel in the sewage works (p.49). The remaining sludge can be sold as a fertilizer.

After the sludge has been removed, the water is still impure. This water is sprayed on to a biological filter. You can recognize this filter at a sewage works by its revolving spray. Bacteria in the filter beds feed on any organic compounds in the water. Any solids which are made go to the gas digester. The rest of the water is finally strained through a fine filter and put back into a river.

Large amounts of industrial waste water do not go through sewage works, but are put into rivers or into the sea next to the factories. In the U.K. and in most other countries there are laws to make sure that the waste is reasonably safe.

Purifying water—a waste of energy?

All the water which comes to houses in the U.K. is treated and fit for drinking. Most of this water is not drunk. About 30% of drinking water goes straight down the lavatory drain. This is a waste of good water and of the money needed to purify it. Scientists are working on ways to avoid this waste. One of their ideas is to collect waste from washing machines and baths, together with rainwater, inside the house. This water could be used to flush lavatories, saving purified water for drinking.

Hard and Soft Water

Do you find "fur" in your kettle? Is it difficult to make bubbles with ordinary soap in your water? Is there a ring of scum round your bath, even if you think you are quite clean? If so, you live in an area where the water is "hard". Your hot water pipe may look like the one in *figure 11.6*. Over half the water supplies in the U.K. are hard.

Hardness in water

Our water supplies first fall as rain. The rain water runs over rocks and earth before it collects in rivers and reservoirs. Water is a good solvent, so it dissolves many chemicals as it runs over the ground. Some of these chemicals are calcium compounds. It is mainly calcium compounds which cause the hardness of water.

Hard water comes from parts of the country where there is limestone or chalk. Limestone and chalk are both forms of calcium carbonate. Calcium carbonate does not dissolve in pure water, but it does dissolve in rain water. This is because rain water is slightly acidic, since it contains dissolved carbon dioxide. Acids react with carbonates, so the limestone and chalk are slowly dissolved away.

Hardness in water has many disadvantages, including causing blocked pipes, but there are also advantages (*table 11.3*). The water strike of 1983 in the U.K. showed one of these advantages

Figure 11.6 A pipe almost blocked by calcium carbonate scale from hard water.

clearly. Water workers stopped treating soft water to make it harder. This meant that poisonous lead from pipes was dissolved more easily by the water. In some places the level of lead was 100 times more than normal.

Advantages of hard water	*Disadvantages of hard water*
Provides calcium for bones and teeth.	Causes "furring" in kettles, boilers and pipes.
Has a stronger taste which many people prefer.	Wastes soap
Dissolves less lead from pipes.	Produces a scum round baths.
Needed for brewing.	Does not leave clothes so soft after washing.
Probably reduces the chance of getting heart disease.	

Table 11.3 Comparing hard and soft water

Types of hardness

The two main chemicals which cause hardness in water are calcium hydrogencarbonate $Ca(HCO_3)_2$ and calcium sulphate $CaSO_4$.

Water containing calcium hydrogencarbonate can be softened by boiling it. This sort of hardness is called *temporary hardness*. Boiling removes temporary hardness because the calcium hydrogencarbonate is broken down by heating. It forms calcium carbonate:

Calcium hydrogencarbonate ⟶
$Ca(HCO_3)_2$

Calcium carbonate + Carbon dioxide + Water
$CaCO_3$ + CO_2 + H_2O

Calcium carbonate does not dissolve in water, so it forms a solid "fur". This solid is formed whenever the water is heated in kettles, boilers or pipes. To stop the furring, the water must be softened before use.

Water containing calcium sulphate cannot be softened by boiling. This sort of hardness is called *permanent hardness*. It does not cause furring in hot water pipes and tanks, but it does have the other disadvantages of hardness (*table 11.3*).

Removing hardness (water softening)

Some hardness in water is removed in water works. This is done by adding the correct amount of lime. More hardness can be removed in peoples' houses if necessary.

A simple method of softening any water is to add washing soda (sodium carbonate). Washing powders contain this chemical. When the sodium carbonate is added, insoluble calcium carbonate is made:

Calcium sulphate + Sodium carbonate ⟶
 (in hard water)   (washing soda)

$\qquad\qquad\qquad$ Calcium carbonate + Sodium sulphate
$\qquad\qquad\qquad\qquad$ (precipitate)

The calcium carbonate is made as a solid precipitate. It leaves soft water behind, because the calcium compound has been removed from the water. There is more about washing, soaps and detergents on page 198.

A second way of softening water is to use a piece of equipment called a water softener. This is a tube containing a chemical called an ion-exchange resin. The hard water is passed through this tube. Calcium ions, which make the water hard, are exchanged for sodium ions, which do not (*figure 11.7*).

This sort of water softener soon becomes full of calcium ions. It can be cleaned out by pouring salt down it. Salt is used so that the column is refilled with sodium ions.

Soft water is better than hard water for washing and cleaning. However, if you have a water softener it is best not to use it on your drinking water. Remind yourself of the advantages of hard water shown in *table 11.3*.

Questions

1 Explain how caves and gorges are formed in limestone areas.
2 Why do washing powders contain sodium carbonate (washing soda)?
3 Water is a good solvent. Most ionic compounds are soluble in water. There are some simple rules for remembering which common compounds are soluble and which are insoluble in water. Read these rules and use them to answer the questions below.
 (i) All sodium, potassium and ammonium compounds are soluble.
 (ii) All nitrates are soluble.

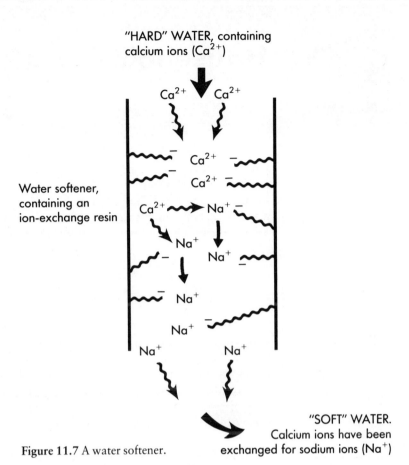

Figure 11.7 A water softener.

(iii) All sulphates are soluble, except lead sulphate, barium sulphate, and calcium sulphate, which is slightly soluble.
(iv) All chlorides are soluble except lead chloride and silver chloride.
(v) All carbonates are insoluble, except sodium carbonate, potassium carbonate and ammonium carbonate.

Which of the following are soluble in water?
(a) Calcium chloride
(b) Calcium carbonate
(c) Ammonium sulphate
(d) Potassium nitrate
(e) Lead carbonate
(f) Copper sulphate
(g) Barium sulphate
(h) Sodium hydrogencarbonate
(i) Magnesium sulphate
(j) Ammonium phosphate.

4   A cold water tank in a house has been overflowing for several weeks across a hot roof. The roof is streaked with a white chemical where the water has flowed across. What is the white chemical likely to be and why is it there?

12 Household Chemistry

Chemistry is not something which happens just in big factories or in school laboratories. It is all around you at home. Cooking a meal, washing the dishes, cleaning the bath, brushing your teeth and taking some medicine are all examples of chemistry in use. Even your house itself has been made from chemicals.

Building the House

Bricks, stone, mortar and cement are the basic building materials. Silicon dioxide, silicates and calcium carbonates are the chemicals which make up these materials. Compounds of silicon and oxygen, mainly silicon dioxide and the silicates, form about 75% of the earth's crust. There is also plenty of calcium carbonate, in the form of limestone, chalk and marble.

Stone and bricks

Stone houses are common in many older towns and villages, close to supplies of local stone. There are about 200 quarries for different building stones in the U.K. What type of stone is used in your area?

Limestone (calcium carbonate) is the most important building stone. Limestones are quite easy to cut and shape. They weather well, and they give a pleasant finish to a building.

Sandstones come next in importance. These are made of silicon dioxide, known as silica. Sandstones are colourful and hard, though they can be cut with diamond-tipped saws.

Granites and slate are silicates. They are not found as commonly as limestone or sandstone in the U.K., but they are much used in their own areas.

Bricks do not occur naturally. They are made from clays, which are also silicates. The clay is ground up and mixed with water. It is then shaped to make a "green" brick and the brick is baked in a kiln.

Mortar and cement

Limestone and clay together are used for making mortar and cement. The two materials are crushed and mixed with water to form a slurry. The slurry is heated in a rotating kiln at about 1500°C, so that it is chemically changed. After cooling, the mixture is crushed and mixed with gypsum (calcium sulphate) to make

Portland Cement. Water and sand are added to the cement in the right amounts to make mortar. Concrete can be made from the mortar simply by adding gravel or other small stones.

The complete house

Many other materials go into building the final house, from glass to the modern insulation materials made from coal and oil. Study the picture of the "energy-saving house" (*figure 12.1*), which illustrates some of these materials.

Figure 12.1 The Low-Energy House Laboratories of the Building Research Establishment.

The house on the left is the Heat Pump House. The blackened aluminium roof collects solar radiation to heat the house. The ground floor is insulated with foamed polystyrene and the walls with glass fibre. The roof is insulated with fibreglass between the joists.

The three terraced houses are built of similar materials but heated in different ways. The house on the right is the Solar House, with solar panels on its roof.

Cooking

Cooking is a chemical reaction. The chemicals which make up food are broken down by heating them with water. This makes them easier for us to digest and it often makes them taste better.

Like all chemical reactions, cooking can be speeded up. This saves us time and saves fuel. Imagine a large potato. If you put it into a pan of boiling water it will take a long time to cook. You can cook it faster by making the temperature higher using a pressure cooker, or by cutting it up into small pieces.

Pressure cookers

Water normally boils at 100°C, but not inside a pressure cooker. The pressure makes it boil at a higher temperature, often 120°C. Many chemical reactions go twice as fast if the temperature is 10°C higher. If the temperature is 20°C higher, the reaction will go four times as fast. Potatoes which normally cook in 20 minutes will only take about 5 minutes in a pressure cooker.

Chopping up food

If a large piece of food is cooked, heat takes a long time to get through the surface and into the middle. Food cooks more quickly if it is chopped up, because there is more surface for the heat to get through (*figure 12.2*).

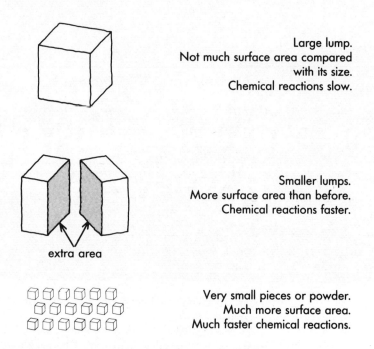

Figure 12.2 Surface area increases when chemicals are cut up.

An increase of the surface area is often used to speed up chemical reactions, apart from its use in cooking. This has its dangers as well. Chemicals which normally burn can even explode if they are in very small pieces. A candle can be used to explode custard powder, which is mostly flour. There have been serious accidents in flour mills when flour dust has exploded. The same thing can happen in coal mines, when a cloud of coal dust explodes.

One way of cooking food quickly without chopping it up is to use a microwave oven. Microwaves can carry heat energy right through the food very quickly. The surface area of the food is not important for microwave cooking.

The Drinks Industry

The drinks industry is big business all over the world. In the U.K. alone, people drink 6500 million litres of beer each year and 400 million litres of wine, in addition to spirits and soft drinks.

Drinks are solutions of different chemicals in water. Alcoholic drinks, including beer, wine and spirits, contain the chemical ethanol. People have known how to make alcoholic drinks for centuries. In small amounts they make pleasant and relaxing drinks. Large amounts can cause great problems. Alcoholism is a serious disease today (p.194).

Beer and wine

Beer and wine are easy to make. Many people make these drinks in their own homes, although they are not allowed to sell the produce. Alcoholic drinks can be made from any fruit or vegetable which contains carbohydrates like starch and sugar.

Beer is usually made from carbohydrates in barley, although home-brewers add extra sugar from a packet. Hops are included to give the beer its bitter taste.

Wine is usually made from grapes. Home wine-makers use many other substances, including elderberries, blackberries, rhubarb and even dandelions to give the wine its flavour.

The carbohydrates are turned into alcohol by a chemical process called **fermentation**. Yeast is needed to carry out these reactions. Yeast is a living organism. It contains proteins called *enzymes* which can convert starch to ethanol. Enzymes are biological catalysts, which need some warmth to work properly. A temperature of about 25°C is often used. It must not be too hot, or else the enzymes are damaged and the yeast is killed.

The starch is first broken down to sugars like glucose. The glucose is then fermented to produce ethanol and carbon dioxide:

$$\text{Starch} \rightarrow \underset{\substack{\text{(a sugar)}\\ C_6H_{12}O_6}}{\text{Glucose}} \rightarrow \underset{\substack{\text{(an alcohol)}\\ 2C_2H_5OH}}{\text{Ethanol}} + \underset{2CO_2}{\text{Carbon dioxide}}$$

On an industrial scale, beer is brewed in enormous vats (*figure 12.3*).

Home brewing and wine-making are much simpler. A diagram of a typical wine-making jar is shown in *figure 12.4*. Notice the air-lock, which allows carbon dioxide to escape but stops air getting back into the jar. Bacteria in the air can turn ethanol into ethanoic acid, which means that the wine turns to vinegar.

Beer contains about 4% alcohol. Chemicals are added to stop the beer going sour like vinegar. Wine contains about 11% alcohol. The yeast is killed if there is much more alcohol than this, so it is not possible to make stronger drinks by fermentation. Some stronger drinks, like sherry and port, are made by adding pure alcohol. Others, including whisky, are made by fractional distillation.

Figure 12.3 Beer brewing in the fermenting vessels at the Alloa Brewery in Scotland.

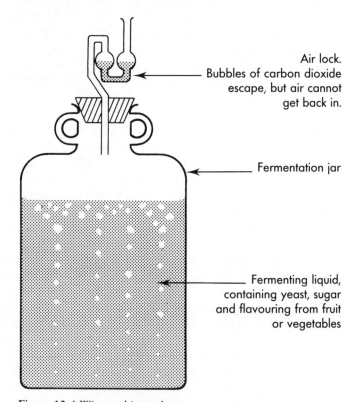

Figure 12.4 Wine making at home.

Spirits

Spirits are made by distilling the solution of the alcohol. This method is used because ethanol, which is the alcohol, has a boiling point of 78°C while water has a boiling point of 100°C. The ethanol can be made more concentrated by fractional distillation. It is illegal to make spirits yourself.

Many spirits are sold as 70° "proof". This does not mean that they contain 70% ethanol. The word "proof" comes from the old way of measuring the amount of alcohol in spirits. The measurement was done by Customs and Excise officers, so that they could tax the drink correctly. What they did was to pour the drink over gunpowder. If the gunpowder could still be lit, the drink was "proof". If not, it was "underproof", because it contained too much water. 70° proof today means about 40% ethanol.

Ethanol is not the only alcohol and it is not just found in alcoholic drinks. **Alcohols** form an important series of organic chemicals. Examples of alcohols and their uses are shown in *figure 12.5*.

ALCOHOL	MOLECULAR FORMULA	STRUCTURAL FORMULA	
Methanol	CH_3OH	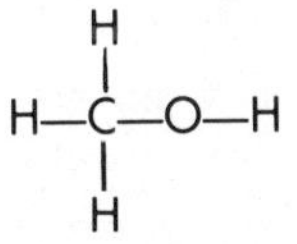 H–C(H)(H)–O–H	Used to make glues, resins and paintstripper. Added to ethanol in industrial spirit and methylated spirits.
Ethanol	C_2H_5OH	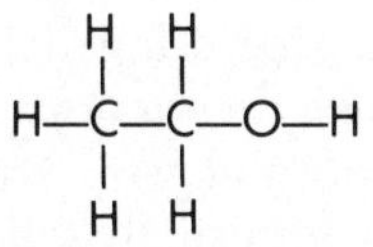 H–C(H)(H)–C(H)(H)–O–H	Made industrially by reacting ethene with steam: $C_2H_4 + H_2O \rightarrow C_2H_5OH$ Used as a solvent in industrial spirit (95% ethanol, 5% methanol). Used as a fuel in methylated spirits (approx. 90% ethanol, 10% methanol with some paraffin oil and a purple dye).
Propanol	C_3H_7OH	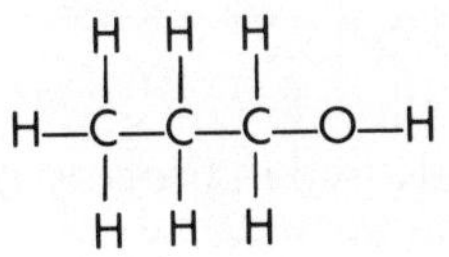 H–C(H)(H)–C(H)(H)–C(H)(H)–O–H	
Butanol	C_4H_9OH	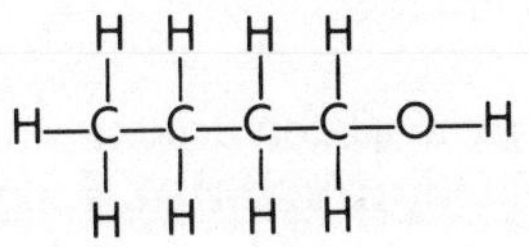 H–C(H)(H)–C(H)(H)–C(H)(H)–C(H)(H)–O–H	

Figure 12.5 Alcohols.

The alcohol problem

Alcohol is a drug. It depresses the nervous system. Even small amounts in the bloodstream, less than the legal limit for drivers, can reduce judgement and skills. About a third of all road accidents are connected with alcohol. In the first year of the "Breathalyser Act" there were 40 000 fewer accidents on the roads in the U.K. and 1152 fewer deaths.

Alcohol is widely used as a social drink. Unfortunately, many people do not know when to stop, or cannot stop. About 1 in 15 drinkers become dependent on this drug. They cannot give it up.

There are about half a million alcoholics in the U.K. Alcoholics are likely to have family problems, work problems and health problems. Cruelty of parents towards children is often linked with alcoholism. Alcoholics are more likely to suffer accidents at work or lose their jobs. They may also suffer from liver disease, heart disease and digestive problems.

Treat alcohol carefully.

Chemicals added to foods and drinks

Look carefully at the labels on cans and packets of food in a supermarket. You will find words like "colouring", "preservative", "antioxidant", "emulsifier" and "flavouring". These are examples of the ways in which chemicals are added to foods to make them last longer, taste better or simply look more interesting.

The chemicals which can be added to foods are controlled by governments. In E.E.C. countries, including the U.K., a series of codes has been brought in. The name of the chemical or its code must be shown on most types of food sold in E.E.C. countries.

Colourings (codes E100–E199)

Colours are added to foods for many reasons. They may restore colours lost during food processing, enhance natural colours, or give colour to foods which do not normally have it. For example, processed peas lose the normal green colour. If a colour is not added back, the peas do not usually sell well. Colourings are usually complicated organic compounds.

The following are just a few examples of available colours:

- E102   Tartrazine yellow
- E110   Orange Yellow S
- E124   Ponceau 4R (red in processed strawberries)
- E131   Patent Blue V
- E142   Green S (green in processed peas)
- E153   Carbon Black

Preservatives (codes E200–E299)

Chemicals have always been used to preserve food. Before the modern chemical industry started, people preserved food by salting it or pickling it in vinegar. The Egyptians and Romans used sulphur dioxide, which we still use today to preserve fruit juices. Preservatives are used to stop bacteria from growing and making the food go bad.

Some common preservatives are:

E211 Sodium benzoate
E220 Sulphur dioxide
E223 Sodium metabisulphite
E236 Methanoic acid
E260 Ethanoic acid (the important chemical in vinegar)
E270 Lactic acid
E280 Propanoic acid
E290 Carbon dioxide

Many preservatives are **organic acids**, or salts of organic acids. Organic acids form the fourth important family of organic chemicals, after alkanes (p.63), alkenes (p.67), and alcohols (p.193).

The organic acids can be made from alcohols by oxidation. Ethanoic acid can be made by the oxidation of ethanol. Bacteria carry out this oxidation when wine is turned to vinegar.

Examples of organic acids and some other information about them are shown in *figure 12.6*.

ACID	MOLECULAR FORMULA	STRUCTURAL FORMULA	
Methanoic acid	HCOOH	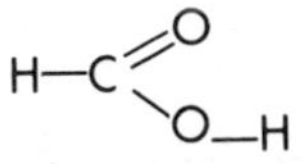 H–C(=O)–O–H	Found in stings from ants
Ethanoic acid	CH_3COOH	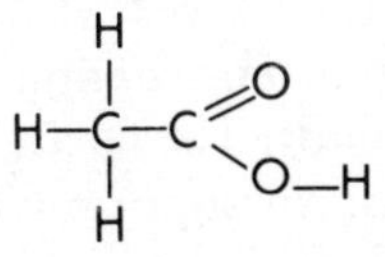 CH_3–C(=O)–O–H	The acidic chemical in vinegar. Used industrially for making some man-made fibres (acetate fibres).
Propanoic acid	C_2H_5COOH	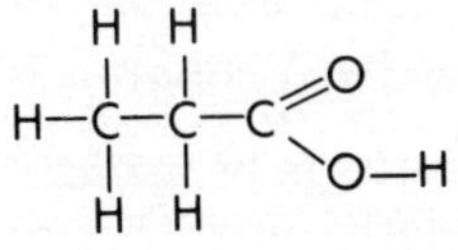 CH_3–CH_2–C(=O)–O–H	
Butanoic acid	C_3H_7COOH	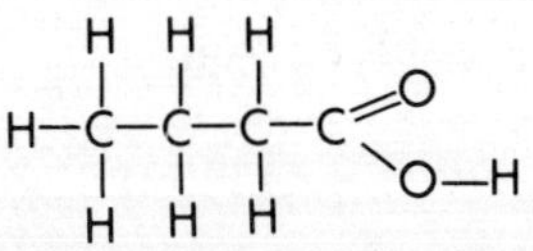 CH_3–CH_2–CH_2–C(=O)–O–H	

Figure 12.6 Organic acids.

Antioxidants (codes E300–E399)

Antioxidants, as the name suggests, are added to stop foods being oxidized. Fats and oils are protected by these chemicals, so that they do not go rancid.

Antioxidants include:

E300 Ascorbic acid (Vitamin C)
E321 BHT (butylatedhydroxytoluene), used in crisps
E330 Citric acid

Emulsifiers, stabilizers and thickeners (codes E400–E499)

This group of chemicals is added to improve the texture of foods, so that the different ingredients will mix together properly.

Chemicals in this group include:

E410 Carob bean gum
E440 Pectin, which helps jams to set
E466 Sodium carboxymethylcellulose

Flavourings

These chemicals do not have codes at present, but they will be found on most food labels. Saccharin, the artificial sweetener, is a chemical in this group.

Other chemicals have a more general use because they bring out the flavours in many different foods. These chemicals are the flavour enhancers. An example is monosodium glutamate which is used particularly in meat and savoury products.

Food scientists experiment with many new chemicals to make new flavours. One group of organic chemicals which are often used are the esters.

Esters

The smell and flavour of many fruits comes from different **esters**, One ester gives the flavour of bananas. Another gives the flavour of grapes, while a third gives the flavour of pears. Other flavours are more complicated. They contain a mixture of esters and other organic compounds. The flavours of plums and figs are two examples of this. An important chemical in these flavours is the simple ester called ethyl ethanoate.

Esters are compounds which can be made from organic acids and alcohols. The general reaction is:

Organic acid + Alcohol → Ester + Water

Ethyl ethanoate can be made from ethanol (the alcohol) and ethanoic acid (the organic acid). The alcohol and the acid are heated with a catalyst of concentrated sulphuric acid:

$$\text{Ethanoic acid} + \text{Ethanol} \xrightarrow[\text{conc. } H_2SO_4]{\text{[heat with]}} \text{Ethyl ethanoate} + \text{Water}$$

$$CH_3COOH + C_2H_5OH \longrightarrow CH_3COOC_2H_5 + H_2O$$

Other esters can be made in the same way (*figure 12.8*).

The ester ethyl ethanoate is not only found in plums and figs. We make it to use as a solvent for glues and nail varnishes.

Figure 12.7 A typical food label.
Find the names of the chemicals added to the drink, using the codes given on the label.

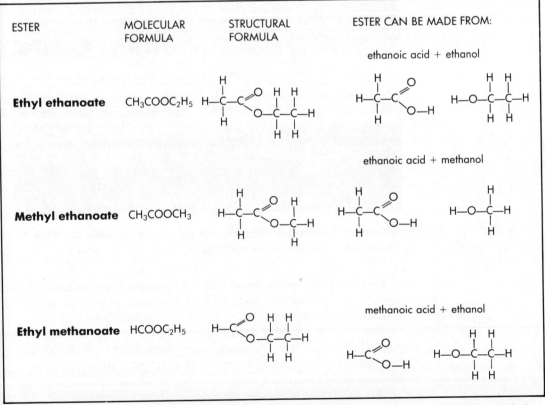

Figure 12.8 Esters. The first half of the ester's name comes from the alcohol, while the second half comes from the acid.

Household Chemistry

"Bicarbonate of soda"

Bicarbonate of soda, which chemists call sodium hydrogencarbonate $NaHCO_3$, is used in baking cakes and some types of bread and biscuits. It is one of the chemicals in baking powder, and it helps to make cakes rise.

When the cake is heated in the oven, the sodium hydrogencarbonate reacts with acids in the cake mixture. Bubbles of carbon dioxide are produced, which make the cake rise. Baking powder contains an acid already mixed in with the sodium hydrogencarbonate.

Soaps and Detergents

Most of our cleaning is done using water, but water on its own is not much good. It may seem strange, but water is not good at wetting many materials or at removing the dirt. Soaps and detergents are added to water so that materials can be cleaned properly (*figure 12.9*).

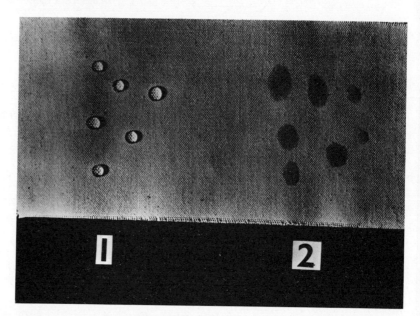

Figure 12.9 Ordinary tap water does not wet unbleached cotton (1). When "Teepol" detergent is used, the water quickly soaks into the fabric (2).

Soaps have been used for at least 2000 years. They are made from animal or vegetable fats and oils. These soaps are not used much today, except for making the bar soaps which you find in kitchens and bathrooms. Detergents made from crude oil are used instead. Detergents are especially useful with hard water because they do not form a scum like ordinary soap. This also means that no detergent is wasted.

Many different types of detergent have now been made by chemists for different purposes. For example, some detergents are good for cleaning glass while others are better for cleaning plastic.

Making soaps

Soaps are salts of organic acids called "*fatty*" *acids*. A typical fatty acid is stearic acid. The soap which is made from it is the salt sodium stearate.

Soaps like sodium stearate are made by boiling animal fats or vegetable oils with an alkali. The tallow from cows and sheep or the oil from palm kernels are most commonly used. They are reacted with sodium hydroxide, which is a cheap alkali.

Animal fats and vegetable oils are esters, which are compounds of alcohols with organic acids like stearic acid. The alcohol is usually glycerol, which is also known as glycerine. When the fat is boiled with a solution of sodium hydroxide, glycerol and the fatty acid salt are produced.

Fat + Sodium hydroxide → Glycerol + Sodium stearate
(ester of   solution   (soap)
glycerol and
stearic acid)

The soap is formed as a precipitate when ordinary salt is added to the mixture. It can then be washed and purified. The glycerol can also be purified. It is a valuable by-product.

Perfume and colouring chemicals can be added to the soap, which is then pressed into bars and sold.

One problem with ordinary soaps is the scum which is formed with hard water. The scum is formed because the soap, sodium stearate, reacts with the calcium compounds in hard water:

Calcium hydro-   Calcium stearate   Sodium
gencarbonate + Sodium stearate →   (insoluble   + hydrogen-
  (soluble)   (soluble)   precipitate)   carbonate
                                                      (soluble)

Calcium stearate is formed as an insoluble scum which is unpleasant and wastes soap.

Making detergents

Detergents, like soaps, are salts of organic acids. Detergents are made from the hydrocarbons in crude oil. The hydrocarbons are reacted with concentrated sulphuric acid, followed by sodium hydroxide. Sodium alkylbenzenesulphonate is a typical detergent made in this way.

About 14% of our sulphuric acid is used for making these detergents. This shows the importance of detergents today. Not only are they used in ordinary washing powders, but also in laundries and dry cleaners, for paper making, for degreasing metal in engineering industries, for cleaning textiles and in agricultural sprays.

A typical washing powder

Only 20% of most washing powders is detergent. Many other chemicals are added to boost their cleaning power and to make

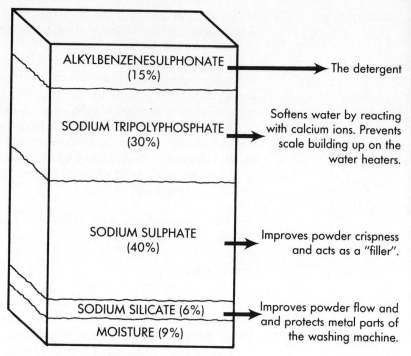

The powder will also contain small amounts of a bleach, a fluorescer to improve the whiteness of cotton and linen, chemicals to stop dirt getting back onto the fabric and perfumes.

Figure 12.10 Chemicals in a typical washing powder. The detergent only makes up 15–20% of most washing powders.

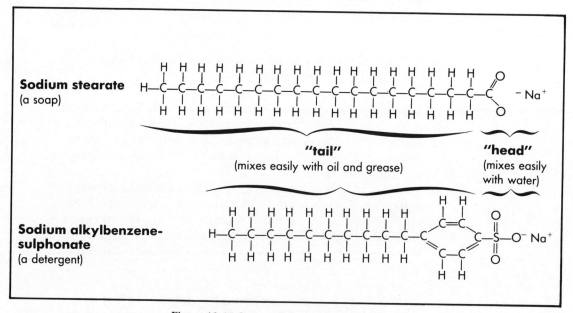

Figure 12.11 Soap and detergent molecules.

200 Chemistry in Use

How soaps and detergents work

them attractive and saleable. The different amounts of some of the chemicals which can be used are shown in *figure 12.10*.

Soaps and detergents are made of the same sort of molecules (*figure 12.11*). The "head" of the molecule is a chemical group which mixes easily with water. The "tail" is a long hydrocarbon chain, like the hydrocarbons in crude oil. This tail buries itself in any dirt or grease while the heads stick out into the water. The dirt is surrounded by detergent molecules and carried into the water, so that it can be washed away (*figure 12.12*).

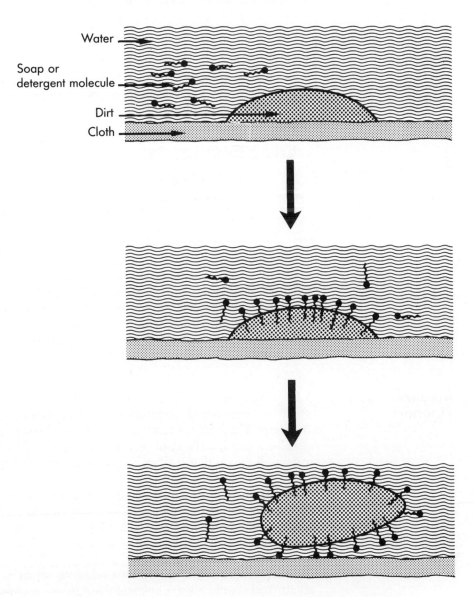

Figure 12.12 Soap or detergent removing dirt from a piece of cloth.

Household Chemistry

The pollution problem

Soaps are made from chemicals found in living animals and vegetables. They can be broken down by bacteria if they are washed into rivers and lakes. This means that they do not cause pollution. They are said to be biodegradable.

The early man-made detergents were not biodegradable. They caused serious pollution on waterways. Large quantities of foam were made, which affected fish and plant life (*figure 12.13*).

Modern detergents are much better, and chemists are working to improve them. One problem left is the phosphate which is used in washing powders. This causes the same sort of pollution as the phosphate fertilizers which drain off farms (p.168).

Bleaches

Many household bleaches contain chlorine, as you can tell from their smell. The active chemical in these bleaches is sodium hypochlorite NaClO. It is made by bubbling chlorine gas through sodium hydroxide solution.

Sodium hydroxide + Chlorine ⟶
2 NaOH + Cl_2

Sodium hypochlorite + Sodium chloride + Water
(bleach)
NaClO + NaCl   + H_2O

This is another example of the importance of salt to us, because sodium hydroxide and chlorine are both made from salt (p.130).

Bleaches are oxidizing agents. Dirt is removed and germs are killed by oxidation. Bleaches are not the only chemicals which remove dirt, but do not try to clean things better by mixing bleaches with other chemicals. Some cleaning chemicals are acids. They will react with bleaches to make poisonous chlorine gas.

Alkaline Cleaners

A large group of cleaning chemicals are alkalis. Common alkalis which are used include sodium hydroxide, sodium carbonate and ammonia. These alkalis react with the fats and oils in dirt to dissolve them. It is the same reaction which is used to make soap (p.199). It means that the fats and oils turn into glycerol and soap, which can be washed away.

The strongest cleaners are scouring powders used for jobs like oven cleaning. They contain an alkali, often sodium hydroxide, together with powdered stone like felspar. This rough powder helps to scrape off the dirt. It also scrapes off your skin if you are not careful.

Other cleaners contain ammonia solution, which is a weak alkali. You should always be careful not to mix ammonia cleaners with bleaches, because poisonous fumes of chloramines are made.

Figure 12.13 Foam pollution.

Chemicals for Body Care

Our bodies are made of chemicals, so we use chemicals to keep them clean and healthy. Advertisements for these chemicals bombard us every day, in the streets and on the TV. We are told to buy this toothpaste, that anti-perspirant, this pill or that deodorant. These chemicals can be divided into two groups. There are the cosmetics, which are used for general care of the body, and the medicines, which are used to cure bodily disorders.

Cosmetics

The term "cosmetics" covers a wide range of chemicals and their uses. Cosmetics generally include skin-care products, perfumes, hair-care products, hand creams, make-up, deodorants, bath products and toothpastes.

A single cosmetic will usually contain many different chemicals. Toothpaste is a good example.

Teeth need cleaning because they become coated with a layer of food and bacteria called plaque. The bacteria produce acids from the food, and the acids slowly rot away your teeth.

The main ingredient in toothpaste is a solid abrasive, often aluminium hydroxide, to rub away the food and bacteria. You cannot tell by looking at it that toothpaste contains an abrasive. It is disguised so that people do not think they are rubbing away the surface of their teeth. The abrasive is a suspension in a liquid like glycerol. A detergent is added to make the toothpaste foam. Flavourings like mint are added for taste. Fluoride is included in some toothpastes, because it helps to stop the decay of teeth. Fluoride is added to the water in some areas, for the same reason (p.218).

Medicines

Before the 20th century, most medicines were based on chemicals extracted from plants. Since 1900, thousands of man-made drugs have been added to these chemicals.

There are now at least 5000 different drugs sold world-wide, although a single drug may have many different trade names. For example, there are about 200 different trade names for the simple aspirin.

Drugs have brought enormous benefits to mankind. Pain-killers can be used to reduce minor aches or, like morphine, to reduce the pain caused by severe injury. Antibiotics like penicillin can be used to fight many diseases carried by bacteria. Tranquillizers like valium, which reduce anxiety, form a much-used group of modern drugs. In addition, there are more powerful drugs used to treat severe mental disorders.

Most medicines, like most cosmetics, are complicated chemicals. One of the simpler ones is aspirin, but even aspirin molecules are quite large and complicated (*figure 12.14*).

Aspirin, made in 1899, was one of the earliest man-made drugs. Almost everyone in a developed country has probably used it as a pain-killer. Like most drugs, it has its disadvantages as well as advantages. Aspirin can cause stomach bleeding. It usually causes little bleeding but some people need hospital treatment after taking the drug. Paracetamol, a similar pain-killer to aspirin, does not have this effect although it can affect the kidneys.

Chemistry for Leisure

Many leisure activities use chemistry in one way or another. Two examples will illustrate this. The first is photography, in which chemistry is directly used. The second is the chemistry of the silicon chip, used in thousands of home computers.

Photography

Chemistry is involved in both stages of photography—taking a picture and developing it.

Photographic films are coated in silver bromide fixed in gelatine. Silver bromide is sensitive to light. Black and white films contain one layer of silver bromide. Colour films have three layers, sensitive to different colours (the three primary colours).

When light falls on the film, the silver bromide is "activated". This, like photosynthesis, is an example of light energy affecting a chemical reaction.

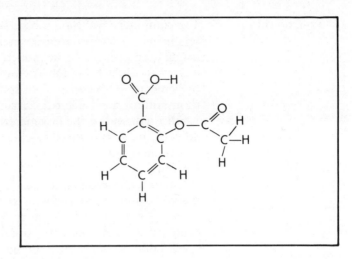

Figure 12.14 An aspirin molecule.

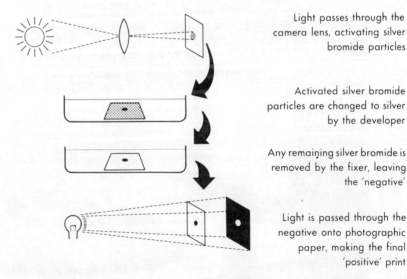

Figure 12.15 Stages in making a photograph.

In the developing stage, the "activated" silver bromide reacts with the developer to make metallic silver. Any unactivated silver bromide can be dissolved away by a chemical called a fixer (often sodium thiosulphate).

After the film has been developed, a negative has been made. Areas of the film exposed to light are now dark (covered in silver), while areas not exposed to light are now light. The negative is then turned into an ordinary print by shining light through it at photographic paper (*figure 12.15*).

The silicon chip

The computer revolution is based on the element silicon, the second most common element in the earth's crust. Silicon is so useful because it is a semiconductor (p.103), and because its electrical conductivity can be controlled by adding small amounts of elements like boron and phosphorus. It is then used to create the thousands of small electrical circuits which make up a silicon chip.

Silicon dioxide is the raw material for making silicon chips. Silicon dioxide is reduced to silicon which is treated further until it is extremely pure. The silicon is melted, crystallized and cut into wafers.

Circuits are printed into the silicon wafers by various chemical methods, until the final silicon chip is made. Chemists have helped in the development of computers by finding ways of carrying out chemical reactions on such a small scale. It is now possible to pack several million electrical components on a chip 5 mm square.

Silicon chips can be mass-produced cheaply. Enormous computing power is now within the reach of businesses and many households (*figure 12.16*).

Figure 12.16 A microcomputer in use. Computers are now a common feature in schools and in many homes.

THE FUTURE

Many chemicals are in short supply in the earth. Metals like copper, zinc and lead, together with the carbon-containing fossil fuels are being used up rapidly. How can we replace these and other metals, or the plastics and energy which are obtained from fossil fuels?

The abundant elements on our planet are iron, aluminium, magnesium, silicon, hydrogen, oxygen and nitrogen. There is plenty of sunlight. It is up to the chemists of the future to find ways of turning these elements into useful materials, perhaps using the energy from sunlight.

Questions

1. Calcium carbonate is important in the building industry, the chemical industry and in agriculture. Give an example of its use in each area. Explain in each case why it is used.
2. Draw a diagram of the equipment you would use to obtain ethanol (alcohol) from a mixture of ethanol and water. Explain how the method works.
3. Orange crystals of potassium dichromate can be used in the breathalyser test for alcohol. Potassium dichromate is an oxidizing agent. It turns green when it reacts with a reducing agent such as ethanol. During the breathalyser test, what new chemical is formed from the ethanol?
4. Explain the difference between a soap and a modern detergent.
5. People today use soap and detergents for cleaning. List the advantages and disadvantages of returning to using soap alone.
6. How do you choose which soaps and detergents you use at home? Ask the rest of your family.

PROJECT 1
[CHAPTER 1]

Lead in the environment

Lead compounds are poisonous. They can affect the body generally, and the brain in particular. Children are affected much more easily than adults. Experiments have shown that people with a higher level of lead in their blood have, on average, a lower IQ ("intelligence"). In other words, the more lead there is in your blood, the less intelligent you are likely to be. This does not prove that lead reduces your intelligence. Perhaps people with less intelligence are more likely to live near places where there is more lead, like main roads. Even so, lead is known to be a brain poison so it probably can affect peoples' intelligence.

Lead compounds can reach humans in many different ways:

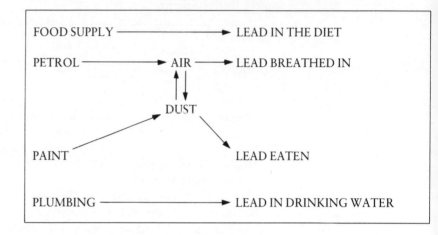

1   *Food supply.* Lead can reach plants either through the soil or through lead-rich dust settling on their leaves. Experiments in 1983 showed that 20% of peeled root vegetables (e.g. carrots) and 34% of washed leaf vegetables (e.g. cabbages) contained more lead than allowed by the U.K. *Lead in Food Regulations.* Humans eat the vegetables and take in the lead.

2   *Petrol.* Lead compounds have been added to petrol in the U.K. for many years. They help the petrol to burn more smoothly. In 1983 between 7500 and 10 000 tonnes of lead came out of car exhausts in the U.K. into the air. This lead can be breathed in by humans or it can fall on crops which are eaten by humans.

Like many other countries, the U.K. is gradually banning the use of lead in petrol because of the worries about lead poisoning.

3   *Paints.* Many paints, especially older ones, contain lead compounds. When the paint flakes off, dust containing lead is made. Old paint is a particular danger in schools because children may be exposed to the lead.

4   *Plumbing.* Lead pipes used to be common, because lead is a

useful metal for plumbing. It does not corrode quickly and pipes can be shaped and fitted quite easily. People now realize the dangers of lead piping, so it is not used. However, many older houses still have lead pipes. If the water stands in the pipes for long, a dangerous concentration of lead can build up. This water should always be run through before drinking any, to make the water supply safe. Many households today have copper pipes for plumbing, but the pipes are often joined by lead solder. The Water Research Centre recommended in 1981 that these solders should be banned. The U.K. government has not banned them at the time of writing (1983).

Questions

1 Summarize the four main ways by which lead can reach humans.
2 Suggest a way of reducing the amount of lead in food.
3 About 90% of lead in the air (U.K., 1983) probably comes from petrol. Suggest as many possible sources of the other 10% as you can.
4 Why is it illegal to paint childrens' toys with leaded paints?
5 Are your water pipes at home made of lead?

PROJECT 2 [CHAPTER 2]

The Oak Ridge Questionnaire

During the Second World War, several thousand scientists worked in special secret laboratories in the United States. They were finding out how to make the atomic bomb. Very few of the scientists really worried about the terrible possibilities of what they were doing. Most of them saw it just as exciting scientific research.

By the summer of 1945 Germany had surrendered, but Japan had not. Many Americans thought that they could end the war with Japan quickly by dropping atomic bombs. At that time, very few people thought that any country except America would be able to make atomic bombs. Also, very few people realized the terrible effects of radiation after a bomb has exploded.

In June 1945, a questionnaire was given to the scientists at the Oak Ridge laboratories. The scientists were asked how the atomic bomb should be used in the war against Japan. They could choose one of five possible answers. Here is a summary of those answers:

1 The atomic bomb should be used in the war to make Japan surrender as soon as possible.
2 A small bomb should be used against Japan. The Japanese should then be told to surrender before a big bomb was dropped.

3 A demonstration of the bomb should be given in a desert in America, with some Japanese watching. They should be told to surrender, or the weapon would be used against them.
4 The atomic bomb should not be used in the war, but a demonstration of its power should be given.
5 The atomic bomb should not be used in the war, and all information about it should be kept as secret as possible.

> **Things to do**
> 1 Which answer would *you* choose?
> 2 Explain the reasons for your choice.
> 3 Find out how many people in the class choose each answer. Compare this with the answers given by the scientists in 1945. These are the percentages of the scientists who chose each answer:
> (1) 15%   (2) 46%   (3) 20%   (4) 11%   (5) 2%

PROJECT
3
[CHAPTER 4]

Anaesthetics

General anaesthetics

Anaesthetics are chemicals which numb a small area of the body or produce a form of sleep. The first group of chemicals are called local anaesthetics. The second group are the general anaesthetics, used for major surgery. Before anaesthetics were discovered, any sort of surgery, from removing a tooth to amputating a limb, was agony for the patient. We can be grateful to chemists that surgery today is nothing like as terrifying.

Most anaesthetics are gases or volatile liquids. The vapours can be breathed in and taken up through the lungs. They travel through the blood to reach the brain, where they cause unconsciousness or "anaesthesia".

The first anaesthetics were found about 100 years ago. They included nitrous oxide N_2O (laughing gas), ether $C_2H_5OC_2H_5$, and chloroform $CHCl_3$ (properly known as trichloromethane).

Nitrous oxide is still used today. Its advantages are that it is not flammable or toxic, but it only causes a light anaesthesia. It is known as laughing gas because people coming round after the anaesthetic sometimes make uncontrollable noises which sound like laughter. Nitrous oxide is part of the "gas and air" mixture given to women to reduce pain in childbirth.

Ether was an important early anaesthetic because it produces deep unconsciousness. Unfortunately it is highly flammable, so it is dangerous to use. Fires could be started just from a spark in the operating theatre. Ether is not used today.

Chloroform gives good anaesthesia and is not flammable. However, it was found to be toxic, causing liver damage in particular. Like ether, it is no longer used.

The ideal anaesthetic needs to produce a deep anaesthesia but not to be toxic or flammable. Scientists started to search for such a compound in 1951. They eventually found a compound which they called halothane:

$$\text{F}_3\text{C}-\text{CHBrCl}$$ 2-bromo-2-chloro-1,1,1-trifluoroethane (Halothane)

This anaesthetic is now widely used in hospitals. It is a volatile liquid which produces deep anaesthesia and is not flammable or toxic.

Local anaesthetics

There is always a risk in using general anaesthetics, so local anaesthetics often have advantages. Most operations at a dentist will be carried out using local anaesthetics.

Cocaine and lignocaine are two chemicals which can be used as local anaesthetics. A third is chloroethane C_2H_5Cl. As it evaporates from the skin it takes heat away from the skin surface. The nerve endings are numbed by the cold, so no pain is felt during an operation.

Questions

1. A new anaesthetic is discovered which is a solid. How could you use it?
2. Describe the disadvantages of each of the three early anaesthetics.
3. Nitrous oxide is the only common gas, apart from oxygen, which relights a glowing splint. Suggest why it does, and what happens to the nitrous oxide during the test.
4. Why is halothane such a good anaesthetic?
5. Write down the structural formula of an isomer of halothane.

PROJECT **4** [CHAPTER 6]

Chemistry in space

Materials for space craft

Space travel has been with us for over 20 years, but the problems have not changed. First of all the most suitable materials must be chosen to build the space craft. Then a fuel is needed to blast the capsule out of the atmosphere. When the craft is in space, it needs energy to work the electrical and life-support systems. The astronauts themselves need food, oxygen and a comfortable environment.

It takes a lot of energy to fire a rocket into space, so the space craft must be as light and strong as possible. Aluminium, magnesium and some titanium alloys are especially used for their lightness, together with strong stainless steels. These metals also have the advantage that they are relatively cheap, easy to work (except titanium), and stable in space.

Some parts of a space craft undergo great heating. Special superalloys containing nickel or cobalt can be used here. The problem is greatest when the capsule re-enters the atmosphere. The temperature can easily reach 2500°C. This would kill the astronauts unless the heat was conducted away. To avoid this, the nose cone can be coated with the plastic Teflon (now used in non-stick pans). Some of the Teflon burns away, absorbing the heat. The rest, because it is plastic, insulates the capsule from the heat.

Rocket fuels

The chemical reaction which powers a rocket is the same as that of any other fuel burning. It is just faster and more violent. All that is needed is the fuel itself and an oxidizer or supply of oxygen. When the fuel burns, gases are forced out of the rocket, so they propel the rocket in the opposite direction.

The American Saturn rocket uses kerosene (jet fuel) with liquid oxygen, and liquid hydrogen with liquid oxygen. Many other fuels have been used, including plastics and rubbers as solid fuels. One unpleasant mixture is hydrazine with red fuming nitric acid as the oxidizer.

Power in space

Solar cells and fuel cells are two important sources of power. Solar cells, made of silicon, convert energy from the sun into electricity. Fuel cells run on hydrogen and oxygen. The gases are reacted together in a special way to make electricity.

Life-support

Astronauts have to create, in space, the environment they would have on earth. Look at what has to be done and remind yourself of what we take for granted here on earth:

Problem	Solution (in the Apollo missions)
CO_2 removal	Lithium hydroxide (this is an alkali; it absorbs the acid CO_2).
O_2 supply	Liquid oxygen store.
Temperature control	Solar powered cooler and water boiler.
Humidity control	Condensation and removal of water.
Contaminant removal	Lithium hydroxide and charcoal (charcoal is good at absorbing many substances).
Water supply	Fuel cell.
Food supply	Freeze-dried dehydrated food.
Personal hygiene	Bactericide—detergent wipes.
Waste management	Defecation glove and urinal.

Questions to think about

1 Why are liquid and not gaseous oxygen and hydrogen used as rocket fuels?
2 Explain how the astronauts' water supply comes from the fuel cell.
3 Why is space food stored in a dehydrated form?
4 The atmosphere in many space craft contains the same amount of oxygen as on earth, but no nitrogen. Suggest a reason for this.
5 Disaster struck the Apollo 13 moon shot when a heater element overheated in a fuel cell. A small spark was made, and the whole fuel cell exploded. Why did it explode?
(The astronauts returned safely to earth.)
6 Why can plastics and rubber be used as rocket fuels?

PROJECT 5
[CHAPTER 7]

Choosing aluminium alloys

Over 200 different forms of aluminium alloys are made. Some information about seven of these alloys is given to you as follows:

1. *Electrical resistance.* The bigger the number, the more the alloy resists an electric current.
2. *Strength.* The bigger the number, the stronger the alloy.
3. *Corrosion resistance.* On a scale from A = very good to E = very bad.
4. *Corrosion resistance under stress.* On the same scale.
5. *Anodizing quality.* On a scale from A = anodizes well to E = anodizes badly.
6. *Price.* Low means about £1000 a tonne. High means about twice that.

Alloy	Electrical resistance	Strength	Corrosion resistance	Corrosion resistance (stressed)	Anodizing quality	Price
6061	3.7	117	B	A	C	Medium
6101	2.9	221	A	A	B	Medium
1350	2.8	83	A	A	A	Low
7001	5.6	255	C	C	D	High
5083	5.9	317	A	B	B	High
2014	5.1	421	D	C	D	High
3003	4.1	131	A	A	D	Low
5005	3.3	200	A	A	A	Medium

Look carefully at the information, and use it to choose the best alloy for the following purposes, giving your reasons:
(a) Building the underwater structure of a drilling-rig.
(b) Making windings inside electrical transformers.
(c) Building an anodized window facing for a large office block.
(d) Making the body of an aircraft which experiences great strains.
(e) Making strong electrical circuit breakers, called busbars, for power stations.

PROJECT 6
[CHAPTER 8]

Chemical warfare

The world spends well over $500 000 million each year on armies and their weapons. A small part of this money is spent on chemical weapons—chemicals, usually gases, which kill people by poisoning. The explosives used in bombs and shells, although they are chemicals, are not normally thought of as chemical weapons.

Chemical weapons have an awesome power. The U.S.A. probably holds enough of them to kill the whole world population, in theory, several thousand times over. A chemical war in Europe could lead to millions of civilian casualties.

Chemical weapons were first used in modern warfare in 1915. The Germans used chlorine gas against the French during the First World War (p.131). Other gases were also used in the First World War. One of these was hydrogen cyanide. Hydrogen cyanide poisons the blood and often stops people breathing within a minute. It is still probably stored by the U.S.S.R. Mustard gas is the most notorious of the early weapons. It attacks moist areas of the body, causing big, painful blisters.

In 1936, the first of the nerve gases was made in Germany. The nerve gases are far more poisonous than the earlier chemical weapons. They affect the nervous system, causing all the body's muscles to contract. Death results from suffocation. The main nerve gases are tabun, sarin and soman. Sarin causes a tightness and aching of the chest, vomiting, cramps and tremors. At a higher dose, severe convulsions set in, followed by collapse, paralysis and death.

A second group of nerve gases, the V agents, was discovered in the U.K. during the 1950s. They are more dangerous than the earlier nerve gases, because they are less volatile. This means that they evaporate more slowly from the skin, so they stay around for a longer time.

In 1925 many countries agreed to sign the Geneva Protocol, declaring that they would not use chemical weapons in war. About 100 countries have now signed the agreement. Even so, countries can still store chemical weapons. France, the U.S.A. and the U.S.S.R. have the biggest stockpiles. The U.K. has made no new gases since the 1950s. The U.S. stopped in 1969 after the "great sheep massacre" when 6400 sheep died 30 miles away from a laboratory making VX nerve gas. However, the U.S. government decided in 1983 that it would like to make new chemical weapons. The U.S.S.R. has probably not made new weapons since 1971, although no-one in the West is certain.

There is still no ban on making and keeping chemical weapons, although the U.K. tried to get a ban in 1976. These lethal chemicals remain as a possible danger to us all.

Questions

1. What is a chemical weapon?
2. What were the earliest chemical weapons?
3. How do nerve gases work?
4. What does "volatile" mean?
5. Why are the V agents more dangerous than the earlier nerve gases?
6. Which countries have the biggest stores of chemical weapons?
7. What is the Geneva Protocol?
8. Do you think that it is right to use any chemical in war, if it will help you to win?

PROJECT 7 [CHAPTER 9]

Spare part surgery

Many parts of the human body can be repaired or replaced by man-made materials. Plastics and metals are the most useful materials. They are chosen for their strength and flexibility. They are also chosen because the body does not reject them in the way that it can reject transplants of natural organs like kidneys.

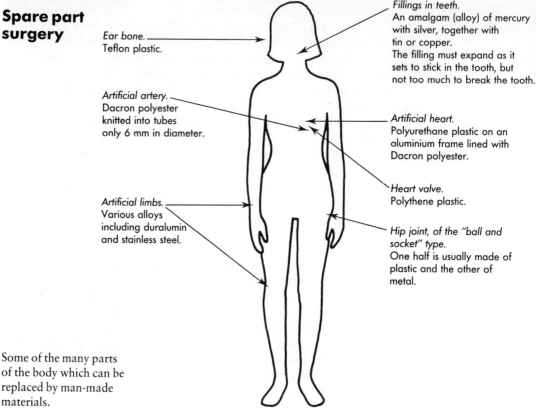

Ear bone.
Teflon plastic.

Fillings in teeth.
An amalgam (alloy) of mercury with silver, together with tin or copper.
The filling must expand as it sets to stick in the tooth, but not too much to break the tooth.

Artificial artery.
Dacron polyester knitted into tubes only 6 mm in diameter.

Artificial heart.
Polyurethane plastic on an aluminium frame lined with Dacron polyester.

Heart valve.
Polythene plastic.

Artificial limbs.
Various alloys including duralumin and stainless steel.

Hip joint, of the "ball and socket" type.
One half is usually made of plastic and the other of metal.

Some of the many parts of the body which can be replaced by man-made materials.

Questions

1. Explain why plastics are chosen to replace parts of the body like heart valves and veins, while metals are used for skull plates and for holding together other broken bones.
2. If you had an artificial heart, what would be your major problems and worries?
3. An artificial heart could cost more than £20 000. Do you think that this is a good way of spending taxpayers' money?
4. If the National Health Service can only afford a few artificial hearts, what sort of people should be given them?
5. One half of an artificial hip joint is normally made of plastic and the other of metal. It is possible to make both from plastic or both from metal. Suggest why this is not done.

PROJECT 8
[CHAPTER 10]

The Seveso Disaster

On 10 July 1976, there was a small explosion at a chemical factory in the North Italian town of Seveso. A chemical reaction in the factory got out of control. The pressure inside the reactor increased as the reaction became exothermic. A safety valve was blown and the chemicals were released into the air. One of the chemicals released was a substance called dioxin. Only 2 kg of dioxin came out, but dioxin is 150 times as poisonous as cyanide. The dioxin fell south of the factory, over an area inhabited by 2000 people.

The first animal deaths were reported on 15 July. Reports of children suffering from skin rashes came a day later. Over 30 people showed signs of burns and poisoning by 22 July. Even so, local leaders said that there was no cause for panic. This is because they did not know that dioxin had been released. The firm Hofmann la Roche, which owns the factory, had not warned them of the danger.

Dioxin is known to cause severe damage to kidneys, the liver, the stomach, intestines and many other organs. People in the Seveso area showed signs of kidney and liver damage. Many people also suffered from chloracne, a nasty form of acne. Chloracne can disfigure people for up to 15 years.

People only realized after two weeks that dioxin had been released. When they knew this they ordered an evacuation. At first only children were evacuated, but later adults were evacuated as well. After three weeks, over 2000 people had been evacuated.

Dioxin can cause birth defects, so pregnant women were very worried. Many women wanted to have abortions. However, at the time of the explosion, abortion was not allowed by Italian law. Many Italian women are Catholics. Abortion is also forbidden by the Catholic Church. Eventually, so many people were worried that the Italian Government passed a law which allowed the women to have abortions. The Catholic Church still said that they should not have abortions.

The company which owns the factory never told the local people about the danger from dioxin. Because they did not tell anyone, no safety measures were taken. Workers at the factory wanted to know about the dangers, but the management did not tell them until two weeks after the explosion. By that time it was too late. Many animals died and many people were badly affected by the poison.

The effects of the Seveso disaster are still not over, years after the explosion. Forty-one barrels of the most dangerous waste from the area were destroyed only in 1983. It takes years for a place to recover after this sort of accident.

Questions

1 Why did the pressure inside the reactor increase as the reaction became exothermic?
2 What damage can dioxin cause?
3 Why was nobody evacuated until two weeks after the explosion?
4 Should the Italian women have had abortions? Explain the reasons for your answer.
5 Small amounts of dioxin can be found in the weedkiller 2,4,5-T. This weedkiller is one of the best killers of tough plants like brambles. Do you think it should be banned?
6 The Seveso factory was making trichlorophenol. Trichlorophenol is needed to make another chemical which is added to soaps, shampoos and toothpastes to kill germs. Is it worth making this chemical if there is a danger of the factory releasing dioxin by accident?
7 What problems do the police and fire-brigade face during a disaster?
8 If a factory exploded in your area would you go and look?
9 Do you think that the media pay too much attention to disasters?
10 Imagine that you are in charge of a small town when a factory explodes. What would you do? Describe your actions step by step.

PROJECT 9
[CHAPTER 11]

Fluoridation of water supplies

By the age of 15, children in the U.K. have, on average, 10 out of 28 permanent teeth decayed, missing or filled.

Nearly one in every three adults aged 16 and over has no natural teeth.

About 7000 sets of dentures are supplied each year to school children.

This is a waste of healthy teeth. It causes pain and worry to many people, especially young children. It is expensive, because our taxes pay for the treatment.

What can be done about it?

One answer is to "fluoridate" the water supply. This can be done by adding small amounts of the compound sodium fluoride to the water. About 5 million people drink fluoridated water in the U.K. Are you on the map?

The "fluoride map" of the U.K. (1982).

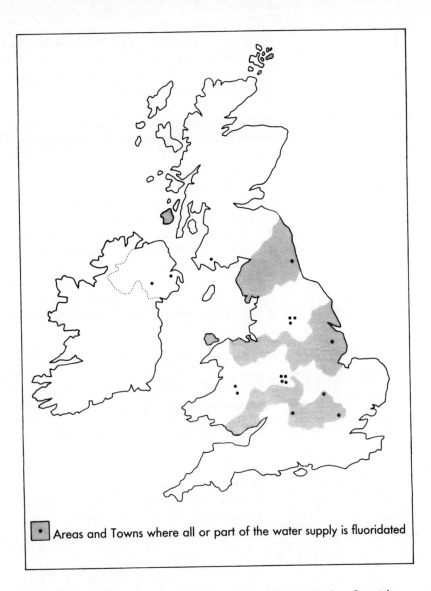

Areas and Towns where all or part of the water supply is fluoridated

Fluoride and tooth decay

Hundreds of studies, from all over the world, show that fluoride does reduce tooth decay. Here are just two examples:

1. A survey in 1979 compared Droitwich (fluoridated) with Hereford (not fluoridated). Every five-year old in Droitwich had just one tooth decayed on average. In Hereford, each child had four decayed teeth.
2. During the ten years after fluoridation in Birmingham, the number of emergency visits to dentists by children dropped from 10 500 each year to 1500.

Problems with fluoride?

Fluoride, like most chemicals, can be harmful in large quantities. It is added to water in small amounts—one milligram per litre. Some natural water supplies have this amount of fluoride anyway.

Hartlepool is one example. There, 51% of five-year olds have no tooth decay. Less than 25% of five-year olds in nearby York (not fluoridated) have no decay. Again it seems that fluoridated water reduces tooth decay.

There is no evidence that fluoridated water causes any harm at all. It is possible to poison yourself with fluoride, but not with fluoridated water. You would have to drink 2500 litres in one go to give yourself a fatal dose.

The right to choose

When the water supply is fluoridated, everyone is forced to drink it. Some people object to this, even though fluoride reduces tooth decay and almost certainly causes no harm to anyone. They say that they should be able to choose for themselves.

Fluoride is particularly useful for children. People who object to fluoridation say that it is up to the parents of a child to give fluoride. This can be done by using fluoride drops brought from a chemist. However, the amount of dental decay in children shows that their parents are not taking much action.

Points for discussion

1. Is it better to prevent a problem like dental decay, than to take action when decay has happened? Think of arguments for and against your point of view.
2. Do you think that fluoride should be added to water if it can reduce tooth decay? Explain your reasons.
3. During a national survey in the U.K. during 1980, about 2000 people were asked Question 2.
    66% answered Yes, 16% answered No, and 18% answered Don't know. How would your class vote?

PROJECT 10
[CHAPTER 12]

Common chemicals

Common name	Proper name	Formula
Alcohol	Ethanol	C_2H_5OH
Baking soda (bicarbonate of soda)	Sodium hydrogen-carbonate	$NaHCO_3$
Calor gas	Butane	C_4H_{10}
Caustic soda	Sodium hydroxide	$NaOH$
Chalk	Calcium carbonate	$CaCO_3$
Epsom salts	Magnesium sulphate	$MgSO_4 \cdot 7H_2O$
Lime (quicklime)	Calcium oxide	CaO
Limestone	Calcium carbonate	$CaCO_3$
Marble	Calcium carbonate	$CaCO_3$
Milk of Magnesia	Magnesium oxide	MgO
Natural gas	Methane	CH_4
Plaster of Paris	Calcium sulphate	$CaSO_4 \cdot \frac{1}{2}H_2O$
Salt	Sodium chloride	$NaCl$
Sand	Silicon dioxide	SiO_2
Slaked lime	Calcium hydroxide	$Ca(OH)_2$
Vinegar	Ethanoic acid solution	CH_3COOH
Washing soda	Sodium carbonate	$Na_2CO_3 \cdot 10H_2O$

1 *Acids and bases* Bases are metal oxides or metal hydroxides. Acids react with bases to form salts. For example, hydrochloric acid reacts to form chlorides, while ethanoic acid reacts to form ethanoates.

> (a) Which of the chemicals above are bases?
> (b) Which of these would react with hydrochloric acid to form common salt?
> (c) What chemicals would be made by reacting vinegar with caustic soda?
> (d) The stomach contains hydrochloric acid. How do you think Milk of Magnesia works to settle stomach complaints?
> (e) Non-metal oxides are acidic. Which of the chemicals above is an acidic oxide?

2 *Carbonates* Carbonates and hydrogencarbonates react with acids to give carbon dioxide. All hydrogencarbonates break down on heating to give carbon dioxide. So do all carbonates except for potassium carbonate and sodium carbonate.

> (a) What do chalk, limestone and marble have in common?
> (b) How does baking soda help cakes to rise?
> (c) What would you see if vinegar was dripped onto chalk?
> (d) How can lime be made from limestone?

3 *Organic*

(a) Which four substances are organic chemicals?
(b) Which of these are used as fuels?
(c) Esters are made by reacting organic acids with alcohols. Two of the chemicals could be reacted together to make an ester used in nail varnish remover. Which two?
(d) Which organic chemical could be used to descale a kettle (scale is calcium carbonate)?

4 *General*

(a) Which of the chemicals can be used for making glass?
(b) Plaster of Paris expands when it sets. Why does this make it useful for taking casts of footprints?
(c) Magnesium sulphate is soluble in water. How could you make it from Milk of Magnesia and dilute sulphuric acid?
(d) How can a suitable alcohol solution be turned into vinegar?
(e) Phosphate fertilizers can be made from calcium phosphate rock. The calcium phosphate is reacted with sulphuric acid. Phosphoric acid is one product. What other chemical is produced? Why could this cause problems?
(f) Give one important use for each of the 17 chemicals.

Appendix 1 Chemical Formulas

A chemical formula gives useful information about a chemical substance. The symbols in the formula show which elements the substance contains. The numbers in the formula give information about the numbers of different atoms or ions which the substance contains.

Two examples should make this clear:

1. *Ammonia*, which is a covalent compound, has the formula
    $$NH_3$$
This shows that ammonia is a compound of nitrogen N and hydrogen H. Each molecule of ammonia contains 1 nitrogen atom and 3 hydrogen atoms. Notice that the number after the symbol shows how many atoms of that element there are. There is no number after the symbol for nitrogen, which means that there is only 1 atom of nitrogen.

2. *Iron (III) oxide*, which is an ionic compound, has the formula
    $$Fe_2O_3$$
The formula shows that iron (III) oxide is a compound of iron Fe and oxygen O. The numbers in the formula show that there are 2 iron ions for every 3 oxide ions in this chemical.

Formulas of Covalent Substances

There are rules for working out the formulas of covalent substances, but they are not often needed. It is best just to learn the formulas of the common covalent chemicals, because there are not many of them. The formulas are given in Table A.1.

Formulas of Ionic Compounds

In order to work out the formula of an ionic compound, it is necessary to know the formulas of the ions themselves. The formulas of some common ions are given in Table A.2.

Ionic compounds contain electrically charged ions, but the compounds have no electric charge overall. The number of positive charges exactly balances the number of negative charges.

Covalent substance	Formula
Hydrogen	H_2
Nitrogen	N_2
Oxygen	O_2
Chlorine	Cl_2
Bromine	Br_2
Iodine	I_2
Carbon monoxide	CO
Carbon dioxide	CO_2
Silicon dioxide	SiO_2
Water	H_2O
Nitrogen monoxide	NO
Nitrogen dioxide	NO_2
Ammonia	NH_3
Sulphur dioxide	SO_2
Sulphur trioxide	SO_3
Methane	CH_4
Ethene	C_2H_4
Ethanol	C_2H_5OH
Ethanoic acid	CH_3COOH

Table A.1 The formulas of some common covalent substances.

Positively charged ions		Negatively charged ions	
Ion	Formula	Ion	Formula
Hydrogen	H^+	Hydroxide	OH^-
Sodium	Na^+	Chloride	Cl^-
Potassium	K^+	Bromide	Br^-
Ammonium	NH_4^+	Iodide	I^-
		Nitrate	NO_3^-
		Hydrogencarbonate	HCO_3^-
Magnesium	Mg^{2+}	Oxide	O^{2-}
Calcium	Ca^{2+}	Sulphide	S^{2-}
Barium	Ba^{2+}	Sulphite	SO_3^{2-}
Copper (II)	Cu^{2+}	Sulphate	SO_4^{2-}
Zinc	Zn^{2+}	Carbonate	CO_3^{2-}
Lead	Pb^{2+}		
Iron (II)	Fe^{2+}		
Iron (III)	Fe^{3+}	Phosphate	PO_4^{3-}
Aluminium	Al^{3+}		

Table A.2 The formulas of some common ions.

1. Sodium chloride

Sodium chloride contains sodium ions Na⁺ and chloride ions Cl⁻. The single positive charge on one sodium ion is balanced by the single negative charge on one chloride ion. The formula of sodium chloride is therefore NaCl, to show that one sodium ion goes with every chloride ion.

NaCl

Notice that the electric charges of the ions are left out of the formula. The formula of sodium chloride is not written Na⁺Cl⁻ even though it contains Na⁺ ions and Cl⁻ ions.

2. Sodium carbonate

Sodium carbonate contains sodium ions Na⁺ and carbonate ions CO_3^{2-}. Two sodium ions are needed to balance the charge of one carbonate ion. The formula of sodium carbonate is written Na_2CO_3 to show this.

Na_2CO_3

3. Calcium hydroxide

Calcium hydroxide contains calcium ions Ca^{2+} and hydroxide ions OH⁻. Two hydroxide ions are needed to balance the charge of one calcium ion. The formula of calcium hydroxide is written as $Ca(OH)_2$.

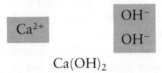

$Ca(OH)_2$

Brackets are needed round the hydroxide ion because the whole of the hydroxide ion is used twice. A group of atoms which is used more than once in a formula must always have brackets round it.

4. Iron (III) oxide

Iron (III) oxide contains iron (III) ions Fe^{3+} and oxide ions O^{2-}. The simplest way to balance the charges with these ions is to have two iron (III) ions with every three oxide ions. The formula becomes Fe_2O_3.

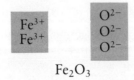

Fe_2O_3

Questions

1. Name the following compounds:

   (a) PbSO$_4$
   (b) Al$_2$O$_3$
   (c) KCl
   (d) NaOH
   (e) CaCO$_3$
   (f) NaHCO$_3$
   (g) (NH$_4$)$_3$PO$_4$
   (h) H$_2$SO$_4$
   (i) MgO
   (j) CuS
   (k) NH$_3$
   (l) CO
   (m) SiO$_2$
   (n) SO$_2$
   (o) CH$_4$.

2. Write the formulas of the following compounds:

   (a) Barium sulphate
   (b) Sodium chloride
   (c) Potassium nitrate
   (d) Sodium carbonate
   (e) Ammonium sulphate
   (f) Calcium oxide
   (g) Calcium hydroxide
   (h) Iron (III) oxide
   (i) Lead carbonate
   (j) Calcium phosphate.

Appendix 2 Chemical Equations and Calculations

Chemical equations, using symbols but not words, have two main uses. Their first use is as a kind of chemical shorthand. Once you have the skill, it is quicker and easier to write equations using symbols than to write word equations. Their second use is for chemical calculations. Using a chemical equation, it is possible to work out how much of one chemical will react with a certain amount of another chemical. Chemists obviously need to know how much of each chemical to use if they are carrying out a chemical reaction.

Chemical Equations

The first stage in writing a chemical equation is to write down a word equation. With more practice, this stage can be missed out.

As an example, when carbon reacts with carbon dioxide (a reaction which takes place in the blast furnace during iron making), the equation is written:

Carbon + Carbon dioxide → Carbon monoxide

The chemicals which react together (the reactants) are written on the left of the arrow. The chemicals which are made (the products) are written on the right.

The next stage is to write down the formula of every substance in the equation:

Carbon + Carbon dioxide → Carbon monoxide
$$C + CO_2 \quad\quad CO$$

The final stage is to "balance" the equation. Atoms cannot be created or destroyed in a chemical reaction. This means that the products must contain exactly the same number of atoms of each element as the reactants.

This is not the case in the equation above. The reactants contain 2 carbon atoms and 2 oxygen atoms in total, whereas the product contains 1 carbon atom and 1 oxygen atom. The equation is balanced by making two molecules of the product, carbon monoxide, as follows:

$$C + CO_2 \longrightarrow 2CO$$
1 atom   1 molecule   2 molecules

The reactants and the products now contain the same number of each type of atom.

The reaction of nitrogen and hydrogen to make ammonia is a more complicated example. The word equation for this reaction is

Nitrogen + Hydrogen → Ammonia

Using the correct formulas, the equation becomes:

$$N_2 + H_2 \rightarrow NH_3$$

The equation is balanced by using 1 molecule of nitrogen to 3 molecules of hydrogen and making 2 molecules of ammonia:

$$N_2 + 3H_2 \rightarrow 2NH_3$$

The equation is balanced because the reactants and the products both contain 2 nitrogen atoms and 6 hydrogen atoms altogether.

Chemical Calculations

Suppose that a chemist needs to know how much lime (calcium oxide) can be made from 100 g of limestone (calcium carbonate).

The equation for the reaction is needed first:

Calcium carbonate → Calcium oxide + Carbon dioxide
$$CaCO_3 \quad\quad CaO + CO_2$$

The equation is already balanced, so there is no need to alter it. The equation shows that 1 formula unit of calcium carbonate gives 1 formula unit of calcium oxide with 1 molecule of carbon dioxide:

$$CaCO_3 \longrightarrow CaO + CO_2$$
1 formula unit   1 formula unit   1 molecule

The term "formula unit" is used instead of "molecule" for calcium carbonate and calcium oxide. This is because they are ionic compounds and are not made up of molecules.

All that the chemist now needs is information about the relative masses of calcium carbonate, calcium oxide and carbon dioxide. These can be worked out from the relative atomic masses of the elements in these compounds. The relative atomic masses of the elements are given in the Periodic Table (p.142).

Relative Formula Mass and Relative Molecular Mass

The relative formula mass (for an ionic compound like calcium carbonate) and the relative molecular mass (for a covalent compound like carbon dioxide) are worked out by adding up the relative atomic masses. The necessary relative atomic masses are:

$$C = 12 \qquad O = 16 \qquad Ca = 40$$

The relative formula mass of calcium carbonate is 100:   calcium oxide is 56:   carbon dioxide is 44:

$$\begin{array}{l} Ca = 40 \\ C = 12 \\ O_3 = \underline{48} \ (3 \times 16) \\ 100 \end{array} \qquad \begin{array}{l} Ca = 40 \\ O = \underline{16} \\ 56 \end{array} \qquad \begin{array}{l} C = 12 \\ O_2 = \underline{32} \\ 44 \end{array}$$

Putting the Relative Masses into the Equation

The relative masses can be written underneath the formulas in the balanced equation:

$$CaCO_3 \rightarrow CaO + CO_2$$
$$ 100 56 + 44$$

The calculation shows that 100 g of calcium carbonate would give 56 g of calcium oxide and 44 g of carbon dioxide.

The same ideas can be used for any other chemical reaction. In this final example, imagine you need to know how much oxygen is used up in burning 16 g of methane.

The balanced equation shows that 1 molecule of methane reacts with 2 molecules of oxygen to give 1 molecule of carbon dioxide and 2 molecules of water:

$$CH_4 + 2O_2 \rightarrow CO_2 + 2H_2O$$
1 molecule   2 molecules   1 molecule   2 molecules

The necessary relative atomic masses are:

$$H = 1 \qquad C = 12 \qquad O = 16$$

The relative molecular mass of methane is 16:

$$\begin{array}{l} C = 12 \\ H_4 = \underline{4} \ (4 \times 1) \\ 16 \end{array}$$

The relative molecular mass of oxygen is 32:

$$O_2 = \underline{\underline{32}} \quad (2 \times 16)$$

The relative masses are written underneath the equation, allowing for the fact that 2 molecules of oxygen are needed to every 1 molecule of methane:

$$CH_4 + 2O_2 \rightarrow CO_2 + 2H_2O$$
$$\;\;16 \quad\;\; 64$$
$$\quad\quad\; (2 \times 32)$$

The calculation shows that 16 g of methane need 64 g of oxygen to burn completely.

Calculations from Formulas

An equation is not needed for all chemical calculations. If the amount of one element contained in a compound is required, only the formula is needed.

The label on a typical bag of urea fertilizer says "46N". This is used to show that the fertilizer contains 46% of nitrogen. The percentage of nitrogen can be worked out from the formula of urea, which is $CO(NH_2)_2$.

The relative formula mass of urea is 60:

$$C = 12$$
$$O = 16$$
$$N = 28 \quad (2 \times 14)$$
$$H = \underline{\;\;4} \quad (4 \times 1)$$
$$\quad\;\; \underline{60}$$

28 out of the 60 units of relative mass come from the nitrogen. The percentage of nitrogen is therefore:

$$\frac{28}{60} \times 100 = 46.7\%$$

The percentage of an element in any compound can be worked out in the same way.

A similar calculation can be used to find out the amount of a metal which could be extracted from a certain amount of its ore. Imagine that you wanted to find out how much iron could be made from 160 g of iron (III) oxide, formula Fe_2O_3.

The relative formula mass of iron (III) oxide is 160:

$$Fe = 112 \quad (2 \times 56)$$
$$O = \underline{\;\;48} \quad (3 \times 16)$$
$$\quad\;\;\; \underline{160}$$

112 out of the 160 units of relative formula mass are iron. This means that 112 g of iron could be made from 160 g of iron (III) oxide.

Questions

1. Balance the following equations. The formulas are all correct and should not be altered. Some of the equations are already balanced.
   (a) $SO_2 + O_2 \rightarrow SO_3$
   (b) $Mg + H_2SO_4 \rightarrow MgSO_4 + H_2$
   (c) $Al + Fe_2O_3 \rightarrow Fe + Al_2O_3$
   (d) $Na + H_2O \rightarrow NaOH + H_2$
   (e) $CaCO_3 + SiO_2 \rightarrow CaSiO_3 + CO_2$
   (f) $CaO + HCl \rightarrow CaCl_2 + H_2O$
   (g) $NaOH + HCl \rightarrow NaCl + H_2O$
   (h) $NH_3 + H_3PO_4 \rightarrow (NH_4)_3PO_4$
   (i) $CH_4 + O_2 \rightarrow CO_2 + H_2O$
   (j) $KClO_3 \rightarrow KCl + O_2$

2. Write balanced symbol equations for the following reactions:
   (a) Zinc carbonate → zinc oxide + carbon dioxide
   (b) Sulphur + oxygen → sulphur dioxide
   (c) Magnesium + oxygen → magnesium oxide
   (d) Sodium carbonate + hydrochloric acid → sodium chloride + carbon dioxide
   (e) Calcium oxide + water → calcium hydroxide
   (f) Iron (III) oxide + carbon monoxide → iron + carbon dioxide
   (g) Iron + copper (II) sulphate → copper + iron (II) sulphate
   (h) Nitrogen monoxide + oxygen → nitrogen dioxide
   (i) Calcium hydrogencarbonate → calcium carbonate + water + carbon dioxide
   (j) Ethanol + oxygen → carbon dioxide + water.

3. Work out the relative formula masses of the substances below. The relative atomic masses are: H = 1, C = 12, N = 14, O = 16, Na = 23, Al = 27, S = 32, Ca = 40, Fe = 56.
   (a) Ammonium nitrate NH_4NO_3
   (b) Calcium carbonate $CaCO_3$
   (c) Ethane C_2H_6
   (d) Sulphur dioxide SO_2
   (e) Sodium hydroxide $NaOH$
   (f) Ammonium sulphate $(NH_4)_2SO_4$
   (g) Aluminium oxide Al_2O_3
   (h) Iron (III) hydroxide $Fe(OH)_3$
   (i) Aluminium sulphate $Al_2(SO_4)_3$
   (j) Octane C_8H_{18}

4. Work out the percentage of nitrogen by mass in
   (a) Ammonium nitrate NH_4NO_3
   (b) Ammonium sulphate $(NH_4)_2SO_4$
   The relative atomic masses are: H = 1, N = 14, O = 16, S = 32.
   Explain carefully why ammonium nitrate has largely replaced ammonium sulphate as a fertilizer.

5. Calcium carbonate breaks down on heating to give calcium oxide and carbon dioxide:
   $$CaCO_3 \rightarrow CaO + CO_2$$
   (a) What mass of calcium oxide would be made from 10 g of calcium carbonate?
   (b) What mass of carbon dioxide would be made at the same time?
   (c) What mass of calcium carbonate would be needed to make 112 g of calcium oxide?
   (d) What mass of calcium oxide would be made from 1 kg of calcium carbonate?
   (e) What is the percentage by mass of calcium in calcium carbonate?
   (C = 12, O = 16, Ca = 40).

Dictionary of 100 Chemical Terms

Acids   Chemicals which contain hydrogen that can be replaced by a metal. Acids can be neutralized by bases, alkalis, metals and carbonates to form salts.
Addition polymerization   Formation of polymers by addition of simple molecules to each other.
Alcohols   A family of organic chemicals containing the −OH group.
Alkalis   Bases which are soluble in water. Alkalis neutralize acids to form salts.
Alkanes   A family of organic chemicals. Alkanes are hydrocarbons which contain only single bonds between carbon atoms.
Alkenes   A family of organic chemicals. Alkenes are hydrocarbons which contain double bonds between carbon atoms.
Allotropes   Different crystalline forms of the same element. Diamond and graphite are allotropes of carbon.
Alloy   A mixture of metals.
Alpha particle   A particle given off by the nucleus of a radioactive chemical. An alpha particle is a helium nucleus, containing two protons and two neutrons.
Amalgam   An alloy of mercury with another metal.
Anode   The positive electrode in electrolysis.
Anodizing   A method used to increase the oxide layer on the surface of a piece of aluminium.
Atom   The smallest part of an element which can take part in a chemical reaction.
Atomic number   The number of protons in the nucleus of each atom of an element.
Bases   Chemicals which react with acids to form salts and water only.
Beta particle   An electron given off by the nucleus of a radioactive chemical.
Biodegradable   Can be broken down by living organisms, usually bacteria.
Boiling point   The temperature at which a liquid boils.
Catalyst   A substance which speeds up or slows down a chemical reaction without being used up itself.
Cathode   The negative electrode in electrolysis.
Chromatography   A method of separating substances, which are often coloured. The chemicals are separated as they move across a material such as paper.
Combustion   Burning.

Compound A chemical formed from two or more elements combined together.
Condensation The change of state from gas or vapour to liquid.
Conductor A substance which allows electricity (electrical conductor) or heat (heat conductor) to pass.
Corrosion The wearing away of a metal by air, water or acids. Rusting of iron is an example of corrosion.
Covalent bond The force holding two atoms together in a molecule. Covalent bonds are formed by the sharing of electrons.
Cracking The breaking up of large molecules into smaller ones. The cracking of alkanes in crude oil to form alkenes is an important example.
Crystal A solid containing atoms, molecules or ions arranged in a regular way.
Diffusion The movement of particles through gases or solutions to form a uniform mixture.
Displacement reaction A reaction in which an element is displaced from its compound by a more reactive element.
Distillation A process by which a liquid is converted into its vapour by heating, and then condensed back to a liquid. It can be used to separate or purify liquids.
Electrode A conductor through which an electric current enters or leaves an electrolyte.
Electrolysis The decomposition of a chemical by electricity.
Electrolysis cell The container in which electrolysis takes place.
Electrolyte A substance which conducts electricity when liquid or in solution, and is decomposed.
Electron A particle with a single negative charge and negligible mass found in atoms.
Electron shells The regions outside the nucleus of an atom in which electrons are found.
Electroplating Coating a substance with a layer of metal using electrolysis.
Element A substance which cannot be broken down into a simpler substance by chemical methods.
Endothermic Describes a chemical reaction which takes in heat.
Equation A way of describing a chemical reaction, giving the names or formulas of all the chemicals which take part.
Evaporation The change of state from liquid to gas or vapour.
Exothermic Describes a chemical reaction which gives out heat.
Fermentation A process in which complex carbon compounds, often carbohydrates, are converted to simpler compounds by enzymes in yeasts.
Filtration A method of separating solids from liquids by passing the mixture through a material like paper. The liquid which passes through is called the filtrate. The solid which is left behind is called the residue.
Formula A way of giving information about the numbers of different atoms or ions which a substance contains.

Fractional distillation (fractionation) The separation of liquids which have different boiling points by collecting the distillates (fractions) at different temperatures.
Gamma ray A form of radiation, similar to an X-ray, which is given out by a radioactive chemical.
Half-life The time taken for half of a radioactive substance to break down.
Hydrocarbon An organic chemical containing only hydrogen and carbon.
Impure Describes a substance containing two or more different chemicals.
Indicator A chemical which is a different colour in acidic or alkaline solution.
Insulator A substance which does not allow electricity (electrical insulator) or heat (heat insulator) to pass.
Ion An electrically charged atom or group of atoms.
Ionic compound A compound containing ions.
Isomers Compounds with the same formula but which have different arrangements of atoms.
Isotopes Atoms of the same element which have different numbers of neutrons.
Lithosphere The earth's crust.
Mass number The total number of protons and neutrons in the nucleus of an atom.
Melting point The temperature at which a solid melts.
Molecule A small group of atoms joined together.
Neutralization The reaction of an acid with a base or alkali.
Neutron A particle with unit mass and no electrical charge, found in the nucleus of an atom.
Non-conductor A substance which does not conduct electricity (or heat).
Non-electrolyte A substance which cannot be electrolysed.
Nucleus The central part of an atom, containing protons and neutrons.
Ore A compound from which a metal or other useful substance can be made. It usually contains impurities.
Organic acids A family of organic chemicals containing the −COOH group.
Organic chemicals Compounds based on the element carbon. Many of these are found in living organisms.
Oxidation Reaction of a chemical with oxygen.
Oxidizing agent A chemical which oxidizes another chemical.
Photosynthesis The building up of carbon dioxide and water into more complex compounds by green plants, using the energy from sunlight.
pH scale A scale from 1 to 14, measuring the acidity of a solution. pH 7 is neutral. Low numbers show an acidic solution. High numbers show an alkaline solution.

Polymer A large molecule formed by joining together many small molecules.
Polymerization The chemical reaction by which a polymer is formed.
Precipitate A solid formed from a reaction in solution.
Proton A particle with unit mass and a single positive charge, found in the nucleus of an atom.
Pure Containing only one chemical substance.
Radiation A way by which energy can be given out by a chemical.
Radioactive Describes an atom which breaks up, giving out radiation.
Redox reaction A reaction in which reduction and oxidation take place.
Reducing agent A chemical which reduces another chemical.
Reduction Removal of oxygen from a chemical.
Relative atomic mass The average mass of an atom of an element, compared with the mass of an atom of ^{12}C, which has a mass of 12 units exactly.
Relative formula mass The mass of a formula unit of a substance compared with the mass of an atom of ^{12}C, which has a mass of 12 units exactly.
Relative molecular mass The mass of a molecule of a substance compared with the mass of an atom of ^{12}C, which has a mass of 12 units exactly.
Respiration The process by which plants and animals use oxygen to obtain energy from foods.
Salts Compounds formed by the neutralization of acids.
Saturated (*a*) A hydrocarbon containing only single bonds between carbon atoms. (*b*) A solution in which no more solid can be dissolved.
Semiconductor A substance which conducts electricity poorly.
Solute The substance which is dissolved in a solvent to form a solution.
Solution Formed when a solute dissolves in a solvent.
Solvent The substance, often a liquid, which dissolves a solute to form a solution.
Structural formula A diagram showing how atoms are joined together in a molecule.
Sublimation The change of state from solid to gas or vapour.
Thermoplastic polymers Polymers which can be reshaped by heating and melting them.
Thermosetting polymers Polymers which are resistant to heat.
Unsaturated Hydrocarbons containing double bonds between carbon atoms.
Valency A measure of the combining power of an atom or ion. It gives the number of covalent bonds which an atom can form or the number of electric charges on an ion.

Index

The most important reference for each entry is in **bold type**
References for uses of chemicals are shown by (U)

Acid rain **23–5**
Acids **161–2**, 169–70
Activity series **79**, 110, 115
Air **175**
Alcoholic drinks **191–4**
Alcohols 51(U), 193(U)
Alkali metals **143**
Alkalis **162**, 169–70, 202(U)
Alkanes 11–4(U), **63–5**(U)
Alkenes **66–74**(U)
Allotropes **146**
Alloys 78, 84, 100(U), 111, 117(U), 119(U), 131(U), **143–5**
Alpha particle **36**
Aluminium 79, 83, **99–103**(U), 110
Aluminium oxide 100–3, **110**
Amalgam **131**(U)
Ammonia **154–5**(U), 165(U), 202(U)
Ammonium nitrate **160**(U), 166
Ammonium phosphate **165–6**(U)
Anaesthetics **210**
Anode **102–3**
Anodizing **110**
Argon **143**(U)
Aspirin **204–5**
Atomic number **33**
Atoms **31–3**, 59–60

Bacteria **154**, 168, 191
Baking powder **198**(U)
Bases **161–2**
Batteries **114–6**, 118
Bauxite **100–1**
Beer **191**
Beta particle **36**
Bicarbonate of soda **198**
Blast furnace **86–8**, 113–4
Bleaches **202**
Bonding 60–3, **143–6**
Boron **44**(U)
Bordeaux mixture **174**
Brass 112(U), **119**
Bricks **188**
Bronze 79, **119**(U)
Butane **11**(U)

Calcium carbonate 88,
 136(U), 171(U), **184–6**, 188(U)
Calcium hydrogencarbonate **185**
Calcium hydroxide 163, **171**(U)
Calcium oxide **171**(U)
Calcium phosphate **163**(U)
Calcium silicate **88**
Calcium sulphate **185–6**, 188(U)
Calculations **227–9**
Calor gas **11**(U)
Carbohydrates **151–2**, 191
Carbon **57**, 85, 91, 103, 146(U)
Carbon cycle **152–3**
Carbon dioxide 16, 22, 27(U), 44(U), 146, **152–3**, 191
Carbon monoxide **16**
Carbonic acid **23**, 162
Car engine **16**
Castner-Kellner process **130–1**
Catalysts **65**, 93, 136, 152, 156, 164, 191
Cathode **103**
Caustic soda *see* Sodium hydroxide
Cement **188–9**
Chalk **184**
Charcoal **48**
Chlorine 129, **131–4**(U), 183(U), 202(U)
Chromium **93**(U), 110
Clay **188**
Clean Air Act **22**
CND **42**
Coal 3–4, **19–22**(U)
Cobalt **93**(U)
Coke 21, **85–6**(U)
Computers **206**
Conductors **103–4**
Contact process **163–5**
Cooking 48, **189–90**
Copper 79, 82, **119–24**(U)
Copper sulphate **122**, 174(U)
Copper sulphide **120**
Corrosion **95**, 110, 112, 117
Cosmetics **203–4**
Covalent bonds **60–3**
Covalent substances 60, **146**

Cracking 14, **64–6**
Crude oil 2, **9–16, 56–7**, 74(U)
Cryolite **103**

DDT **172–3**
Desalination **180**
Detergents **198–202**
Diamond **146–7**(U)
Dioxin **174**
Displacement reactions **124–5**
Drugs 195, **204**
Dry cleaning **133–4**
Duralumin **100**(U)

Electricity 19, 26, 30, 43, 101, **103–5**, 119, 145
Electrode **102**
Electrolysis 102, **109**, 120, 130
Electrolytes **103–4**, 145
Electron 31, **33**, 59–61, 103, 145
Electron shells **59–61**, 107
Electroplating **122–4**
Endothermic **8**
Energy crisis 9, **47–8**
Enzymes **191**
Equations **226–9**
Esters **196–7**, 199
Ethane **11**, 13, 63
Ethanoic acid 191, **195–7**(U)
Ethanol **191–3**(U)
Ethene **66–8**(U)
Ethylethanoate **196–7**(U)
Exothermic **8**, 155

Fats **151–2**, 199
Fermentation **191–2**
Fertilizers 51, **154–60**, 163–9
Fires **26–8**
Fluoridation 204, **218–20**
Food additives **194–8**
Foods **151–2**
Formulas **223–5**
Fossil fuels **2–26**, 163, 167
Fractional distillation **12–14**, 193
Frasch process **176**
Fungicides **174**

Galvanizing **96–7**

Gamma rays **36**
Geothermal energy **52–3**
Giant structures **145–6**
Glass **136–9**
Glucose 152, **191**
Glues **72**
Glycerol **199**, 204
Gold **78–9**(U), 120
Graphite 44(U), 103(U), 116(U), **146–7**(U)
Greenhouse effect **22**
Groups **143**

Haber process **155–6**
Haematite **84**
Half life **39–41**
Halogens 131, **143**
Hard water **184–6**
Helium **143**(U)
Herbicides **174–5**
Hiroshima **31**, 42
Hydrocarbons **2**, 12, 57
Hydrochloric acid **162**
Hydroelectric power 2, **52**, 101
Hydrogen 52, **134–6**(U)
Hydroxides **162**

Imperial smelting process **114**
Indicators **169**
Insecticides **172–3**
Insulators **103**
Ion exchange **186–7**
Ionic bonds **145**
Ionic substances 104, **145**
Ions **104**, 145, 224
Iron 79, **84–9**(U)
Iron (III) oxide **87**
Isomers **63**
Isotopes **33–5**

Lake Erie **168**
Lead 16, 79, 114, 116–8(U), 185, 208–9
Light **152**, 204
Lime **171**(U), 174(U)
Limekiln **171**
Limestone **85–6**(U), 136(U), **171**(U), 184, 188(U)
Limewater **163**, 171

Magnesium **97**(U), **125–6**(U)

Index 235

Magnesium oxide **162**(U)
Manganese **93**(U)
Mass number **34**
Medicines **204**
Melamine **69**
Mercury **130–1**(U)
Metals **77–83**, 143–5
Metallic bonding **143**
Methylated spirits **193**
Methane 8, **11**, 13, 48–51(U), 155(U)
Methanol **193**(U)
Milk of Magnesia **162**
Minamata **131**
Molecules **59–63**, 146–7
Molybdenum **93**(U)
Mortar **188–9**

Naphtha 13, **64–6**
Natural gas 2, **4–9**(U)
Neon **143**(U)
Neutralization **162**
Neutron 31, **33**, 41
Nickel **120**(U), 136(U)
Niobium **93**(U)
Nitrates **154**, 168
Nitric acid **155–60**(U), 162
Nitrogen **154–5**(U)
Nitrogen cycle **161**
Nitrogen dioxide 24, **157–8**
Nitrogen monoxide **156–7**
Nitrous oxide **210**
Noble gases **143**
Non-electrolytes **103–4**, 146
Non-metals **77**
Norfolk Broads **168**
North Sea Gas **5**(U), 155(U)
Nuclear energy 2, 30, **41–5**, 209
Nuclear power stations **43–5**
Nuclear weapons **41–2**, 45, 209–10
Nucleus **31**
Nylon **70**

Octane **11**(U)
Oil *see* Crude oil
Oil shales **24**
Oleum **165**

Organic acids **195–6**(U)
Organic chemicals **57**
Oxidation **8**, 202
Oxides **23**, 162
Oxygen 8, **80**, 152

Paints **72**, 95(U)
Paracetamol **204**
Paraquat **174–5**
Peat 3, **26**
Penicillin **204**

Periodic table 32–3, 58–60, 76, 106, **141–3**
Periods **141**
Permanent hardness **186**
Pesticides **172–5**
Petrol 13, **15–16**, 64, 117
Petroleum *see* Crude oil
pH scale **169**
Phosphates **163–5**, 168
Phosphoric acid **162**, 165
Phosphorous 89, 154, **163**
Photography **204–5**
Photosynthesis **152**
Plastics **66–72**(U), 103(U)
Platinum **93**(U), 120
Plutonium **45**(U)
Pollution 14–6, **22–5**, 45, 71, 94, 131, 138–9, 168–9, 202
Polymerization **66–8**
Polymers **66–72**
Polystyrene **68**
Polythene **66–8**
Potassium **125**, 154, **166**
Potassium chloride **166**(U)
Primary cells **116**
Propane **11**
Proteins **151–2**
Proton 31, **33**
PVC **68–9**, 132

Quicklime **171**(U)

Radiation **35–6**
Radioactivity **35–41**
Recycling **83**, 138–9
Redox reactions **87**
Reduction **87**
Refinery **12–4**

Relative atomic mass **35**, 228
Relative formula mass **228**
Relative molecular mass **228**
Respiration **151–2**
Rubbers **72**, **103**
Rusting **94–7**

Sacrificial protection **97**
Salt *see* Sodium chloride
Salts **161–3**
Sand **136**
Sandstone 3, **188**
Secondary cells **118**
Sewage treatment 49, **183–4**
Silica **188**
Silicon 80, 89, **103**(U), 206(U)
Silicon dioxide 85, **136**(U), 188(U), 206(U)
Silver **78–9**, 100(U), 120, 205
Silver bromide **204–5**(U)
Slaked lime **171**(U)
Soaps **198–202**
Sodium **125**(U), 129
Sodium alkylbenzenesulphonate **199–200**(U)
Sodium carbonate 129, **136–7**(U), 186(U), 202(U)
Sodium chloride **127–9**(U)
Sodium hydrogencarbonate 27(U), **198**(U)
Sodium hydroxide 129, **134–5**(U), 162, 202(U)
Sodium hypochlorite **202**(U)
Sodium nitrate **154**(U)
Sodium stearate **199–200**(U)
Solar energy 4, **54**
Solder **117**(U)
Solvents **133–4**, 184, 193
Spirits **193–4**
Stainless steel **97**(U)
Starch **152**, 191
Stearic acid **162**, 199

Steel 84, **89–93**(U)
Stone **188**
Structural formula **11**
Structure **143–6**
Sugar **51–2**, 191
Sulphur **72**(U), **163**(U)
Sulphur dioxide 16, **24–5**, 164
Sulphur trioxide **164**
Sulphuric acid **118**(U), 162 **163–5**(U), 199(U)

2,4,5-T **174–5**
Tantalum **93**(U)
Tar sands **24**
Temporary hardness **185**
Terylene **70**
Thermit welding **110**
Thermoplastics **68**
Thermosetting materials **68–9**
Tin **79**, **96**(U), 117(U), 120(U), 137(U)
Titanium **93**(U), 130(U)
Toothpaste **203–4**
Transition metals **142**
Transpiration **178**
Trichloromethane **210**(U)
Tungsten **93**(U)

Uranium 41(U), **44**(U)

Valency **60**
Vanadium **93**(U)
Vanadium (V) oxide **164**(U)
Vinegar **191**
Volta river scheme **100–1**

Washing soda **186**(U)
Water 8, 28, 52, 152, **178–86**, 198
Water cycle **178–9**
Wine **191**
Wood **48**(U)
Wood's metal **118**(U)

Yeast **191**

Zinc **96**(U), **111–6**(U), 120(U)